THE EXXON VALDEZ OIL SPILL
and the National Park Service:
A Report on the Initial Response

prepared by
William S. Hanable
with the assistance of Carol Burkhart

National Park Service
Alaska Region
Anchorage, 1990

CONTENTS

ILLUSTRATIONS

Figures

Maps

ACKNOWLEDGEMENTS

I am deeply indebted to many people for their assistance in
preparing this study. Many are listed in the bibliography.
Special thanks are due to Dave Ames, for the idea of assigning an
historian to this project; to Ray Bane, Anne Castellina, Dave
Liebersbach and others for giving time and attention to the
historical record in the midst of the incident; to Marv Robertson
for initiating me into the workings of an ICT; to Mary Ann Roddy
and Barbara Harger for transforming interview recordings into
transcripts; to Kate Lidfors and Carol Burkhart for thoughtful
review and criticism of the manuscript; and especially to the
latter for assistance in revisions and editing, and the final
assembly of the manuscript, photographs, and maps.

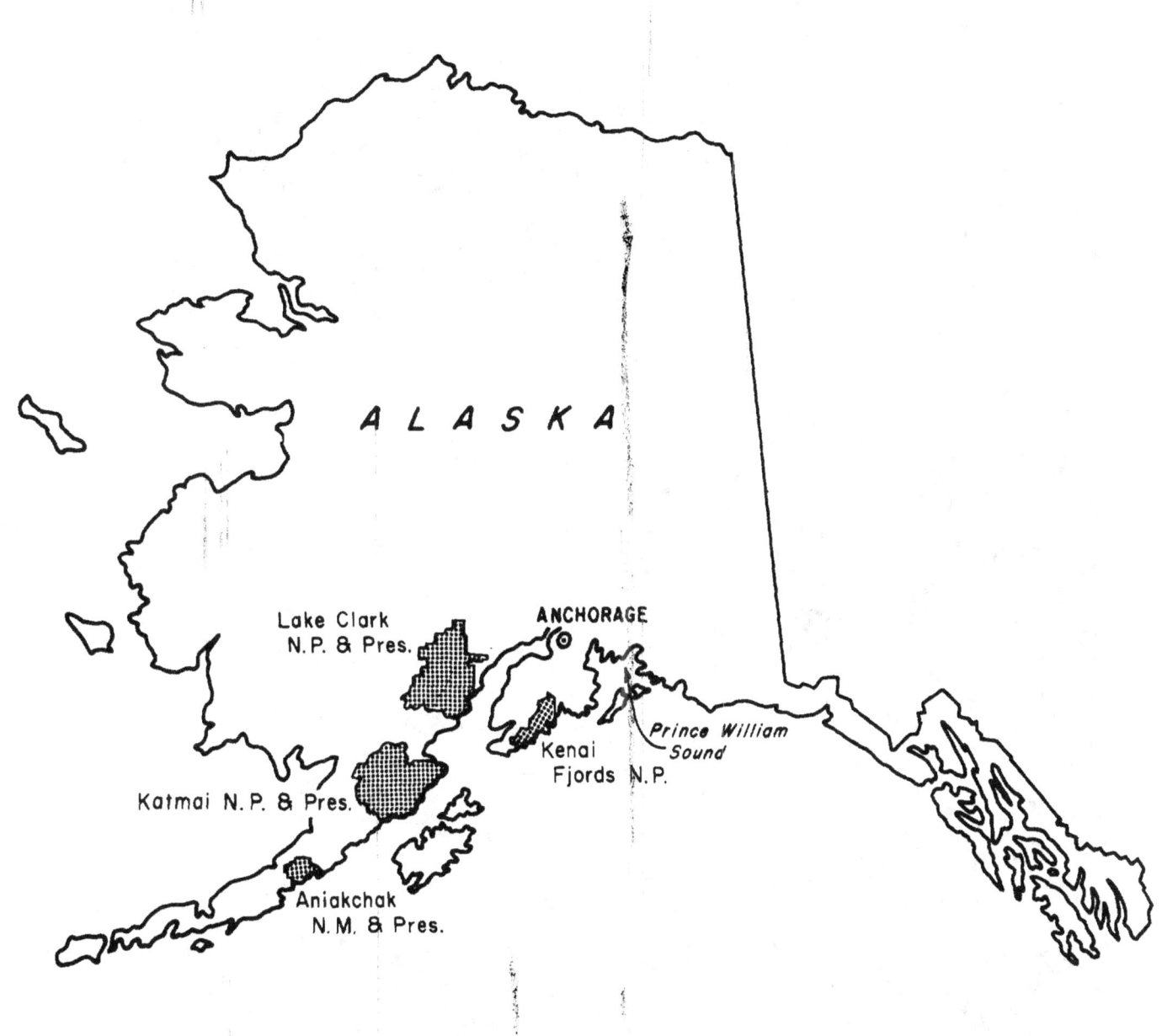

**National Parks in Alaska
Threatened by Oil Spill**

953 | 80,113
FEB 90 | ARO

T/V <u>Exxon Valdez</u> surrounded by boom at Naked Island in Prince William Sound. (Photo courtesy of Karen Jettmar.)

INTRODUCTION

Oil sticks to everything. It is toxic and, like a fire, can kill what it touches. Unlike a fire, which requires a constant combination of fuel, oxygen, and heat, spilled oil is always present and must be physically removed. This comes about either through dispersion by the forces of nature or by the intervention of humans. What follows is the story of how the National Park Service (NPS) responded, in the first several weeks after the disaster, to the largest oil spill to occur in North America.

The initial response reflected the urgent nature of the threat and injuries presented to the land the Service manages for the American people. The NPS will be responding to resource damage, legal ramifications, and other consequences for what may be years to come. Some investigations were only beginning as the first phase of the incident ended. The activities described in this report represent only the first phase.

One federal agency refers to this type of historic narrative as a "Contemporary Historical Examination of Current Operations." Although this contemporary historical examination, underway almost from the beginning of the incident until the first phase ended with the departure of the field teams in the fall, has certain limitations, it also offers particular advantages. Such histories leave records that historians can later reinterpret with the broader perspectives that come with the passage of time. They also can help an organization prepare for future challenges.

The report focuses on the use of the Incident Command System, an existing mechanism for managing federal agency response to fires. The system was applied in Alaska, for the first time, to a different kind of environmental disaster.

The story is complex. There was enough oil so that some remained at its first point of contact while deathly quantities moved to contaminate fresh areas. To make the story more understandable, the following narrative is divided into several parts. Chapter One tells how and where the oil was spilled, provides historical background, identifies NPS resources put at risk by the spill, and describes existing mechanisms for dealing with such threats. It also discusses the initial response of NPS to the spill. Chapter Two elaborates on the methods used to direct and control the NPS response to the spill during the first phase of the spill response, and the interface of that direction and control with similar efforts by other agencies and institutions. Chapter Three is an account of pre-oiling staff and field operations conducted in anticipation of the arrival of the spilled oil. Chapter Four develops the themes of the two previous chapters -- command and control, and field operations -- as they evolved after the oil arrived. Chapter Five both summarizes interpretations of previous chapters and presents additional conclusions.

CHRONOLOGY OF KEY EVENTS

Date	Key Event
Mar 24	Exxon Valdez grounds on Bligh Reef.
Mar 28	Type-I Alaska ICT mobilizes to go to Valdez.
Mar 29	Oil begins to move out of Prince William Sound to flow southwest. Kenai Fjords National Park calls for help in dealing with oil spill. Alaska Regional Office decides to call in ICT to assist Kenai Fjords National Park.
Mar 30	Type-I Alaska has first meeting with Kenai Fjords superintendent and staff. Bud Rice and Page Spencer draft initial plan for pre-oiling assessments.
Mar 31	First ICT pre-oiling assessments begin.
Apr 01	Sen. Stevens encourages ICT work on non-federal lands, and defensive booming. Superintendent Castellina and ICT Commander Liebersbach decide ICT will deploy boom. Multi-Agency Coordinating (MAC) Group formed.
Apr 02	Kenai Peninsula Borough makes arrangements to reimburse NPS for ICT work on non-NPS lands. ICT deploys boom for the first time.
Apr 03	City of Seward joins NPS in unified command of ICT. NPS Tort Team begins establishing chain-of-custody for documentation gathered.
Apr 05	Lake Clark National Park and Preserve requests ICT assistance.
Apr 06	Katmai National Park and Preserve requests ICT assistance. Incident Commander decides to establish branches in Kenai and Homer.
Apr 07	ICT Branches open in Kenai and Homer. "Mini-MAC" established in Homer.
Apr 09	Pre-oiling investigations for Lake Clark begin. Principal defensive booming completed.

Date	Key Event
Apr 10	Oil strikes Kenai Fjords coastline.
Apr 11	Pre-oiling investigations for Kenai Fjords completed.
Apr 12	Patches of oil observed on Katmai beaches.
Apr 13	Kenai Fjords post-oiling assessments begin. Kenai Branch ICT demobilizes.
Apr 14	Exxon assumes responsibility for maintaining boom deployed by ICT.
Apr 15	Pre-oiling investigations for Katmai and Aniakchak begin. Homer ICT Branch demobilizes.
Apr 16	Type-II ICT at Kodiak begins to assist Katmai and Aniakchak.
Apr 19	Type-I ICT at Seward demobilizes.
Apr 26	First major oil strike on Katmai coastline at Cape Douglas.
Apr 29	Pre-oiling investigations for Katmai and Aniakchak are completed.
May 05	Exxon begins removing oil from Kenai Fjords beaches.
May 10	Exxon begins removing oil from Katmai beaches.
May 11	Alaska Regional Office establishes Area Command ICT.
May 16	Type-II ICT mobilizes at Seward.
Jul 02	Oil documented as reaching Aniakchak coast.
Jul 04	Exxon crews begin removing oil from Aniakchak beaches.
Sep 15	Exxon ends cleanup activities.
Sep 30	Area Command, Katmai, and Kenai Fjords ICT demobilized.

Figure I-1

THE ALASKA OIL SPILL
March 24 to June 30

Cumulative extent of ADEC oil spill data

National Parks, Forests, and Wildlife Refuges

Alaska State Parks, Game Refuges, and Critical Habitats

Other uplands

State of Alaska
Dept. of Natural Resources

953 | 80,116
FEB 90 | ARO

CHAPTER 1 - BACKGROUND AND INITIAL RESPONSE

- Background
- Oil Development and Transportation
- Oceanography and Geography
- "A Dragon's Breath of Swirling Death"
- Resources at Risk
- National and Regional Response Mechanisms
- Washington Reaction to ICT Help
- Decision to Call in the Alaska ICT
- Comment

Background

The Tanker Vessel (T/V) Exxon Valdez ran aground on Bligh Reef in Prince William Sound, Alaska, at four minutes past midnight on March 24, 1989. The two-year-old tanker, 987 feet long and 166 feet wide, carried over 53 million gallons of oil destined for Long Beach, California.

The grounding of the Exxon Valdez ruptured eight of the tanker's 11 cargo tanks. Within five hours, over 10.1 million gallons of oil leaked from the tanks.[1] This created the largest oil spill in American history. As spilled oil moved with current and winds out of Prince William Sound to the west, it threatened four National Park units: Aniakchak, Katmai, Kenai Fjords, and Lake Clark.

Oil Development and Transportation

Natives and explorers knew of the presence of petroleum deposits on Alaska's Arctic Coast for many years, but oil seekers did not drill the first well there until 1963. Commercial production did not begin until the early 1970s. A decision to move the product from Arctic Alaska's oil fields to market via an overland pipeline to an ice-free tidewater port and then by tanker created the potential for a devastating oil spill in coastal waters off southcentral Alaska.

A spill at sea in early 1970 demonstrated that potential. An unknown source, believed to be a tanker, discharged dirty ballast or slop oils off southcentral Alaska. The oil first appeared in January. By March, the oil had dappled 1,000 miles of coastline from Montague Island, at the entrance to Prince William Sound, to Shelikof Strait between Kodiak Island and the Alaska Peninsula. Contamination reached Gore Point, in what later became Kenai Fjords National Park, and Swikshak Bay in Katmai National Monument. Officials estimated that the oil killed 10,000 seabirds. They believed that at least 500 marine mammals encountered the oil.[2] This spill of a relatively small

CHRONOLOGY OF OIL DEVELOPMENT IN ALASKA

Year	Event
1885	U.S. Navy Exploring Expedition brings back oil samples from Colville River region
1902	Alaska Development Company brings in well at Katalla
1923	Naval Petroleum Reserve No. 4 established in Arctic Alaska
1945	First test well driven in Naval Petroleum Reserve No. 4
1957	Test well on Kenai Peninsula strikes oil
1968	Test well at Prudhoe Bay strikes oil, taps estimated 10-billion barrel reservoir
1969	State of Alaska sells oil leases for 179 tracts in Arctic Alaska
1970	Eight major oil companies form Alyeska Pipeline Service Company to build and operate pipeline and marine terminal
1974	Congress approves plans for pipeline and marine terminal
1977	First oil flows through Trans Alaska Pipeline and is shipped through marine terminal
1981	New oil field at Kaparuk River, 40 miles west of Prudhoe Bay adds its production to oil flowing through Trans Alaska Pipeline
1981	Alyeska disbands its full-time oil spill response team
1989	T/V _Exxon Valdez_ spills 11 million gallons of oil in Gulf of Alaska in March
	Oil spilled from T/V _Exxon Valdez_ hits Kenai Fjords National Park, Katmai National Park in April and later Aniakchak National Monument

Figure 1-1[3]

amount of oil, probably 3,000 to 6,000 gallons, presaged what might happen if a tanker's hull ruptured.

Valdez, Alaska, became the terminus of the Trans Alaska Pipeline and the location of a marine terminal despite objections of environmentalists and the fishing industry. From the terminal, hoses transfer pipeline oil to tanker vessels. The tankers sail from Valdez through the constricted waters of Prince William Sound. Once outside the sound, the gigantic ships go either to ports on the West Coast of the United States or to the Isthmus of Panama. Some of the Alaskan oil, transported across Panama by pipeline and once again placed in tankers, goes to American ports on the Gulf of Mexico.

Congress rejected the alternative of an all-land pipeline over American and Canadian territory to approve construction of the Trans Alaska Pipeline and marine terminal in 1974. The terminal is on the south shore of Port Valdez, the innermost portion of a fjord known as Valdez Arm. The arm extends northward from the northwest corner of Prince William Sound between Glacier and Bligh islands. Prince William Sound, to which Port Valdez is connected by Valdez Arm, is over 1,000 nautical miles northwest of Seattle. The sound extends east to west from Point Whitshed to Cape Puget, a distance of 150 miles, and south to north from Hinchinbrook Entrance to College Fjord, about 100 miles. On the seaward side, the Gulf of Alaska borders the sound.

The first oil flowed through the Trans Alaska Pipeline in June 1977. Tanker shipments from the terminal at Port Valdez followed immediately. By the 1980s, over 20 percent of the oil supply of the United States flowed through the terminal at Valdez. In 1988 alone, Alaskan fields produced 730 million barrels of oil, most from the North Slope. The bulk of this oil went to market through the Trans Alaska Pipeline and its tanker vessel connection.[4]

Oceanography and Geography

The route of the tankers takes them out of Port Valdez via Valdez Narrows, through Prince William Sound, and into the Gulf of Alaska. In traveling these waters, the tankers slice through the Alaska Coastal Current. The current is a 10-mile-wide flow of water moving westward along the 850-mile gulf coast at a speed of about two knots and carrying about 30,000 cubic yards of water per second.[5]

An offshoot of the current surges into Prince William Sound through Hinchinbrook Entrance. Then it sweeps the mainland and island coasts of the sound, before pouring through Montague Strait to rejoin the main Alaska Coastal Current on its westward journey. Once past Prince William Sound, the coastal current heads southwest along Blying Sound, past Resurrection Bay and Kenai Fjords National Park. At the tip of the Kenai Peninsula, a portion of the current swirls counter-clockwise through Cook

Inlet as far as the Kenai Forelands. Once flushed from the inlet, it rejoins the main stream that carries it down the Shelikof Straits along the west shore of the Alaska Peninsula to Dutch Harbor.[6]

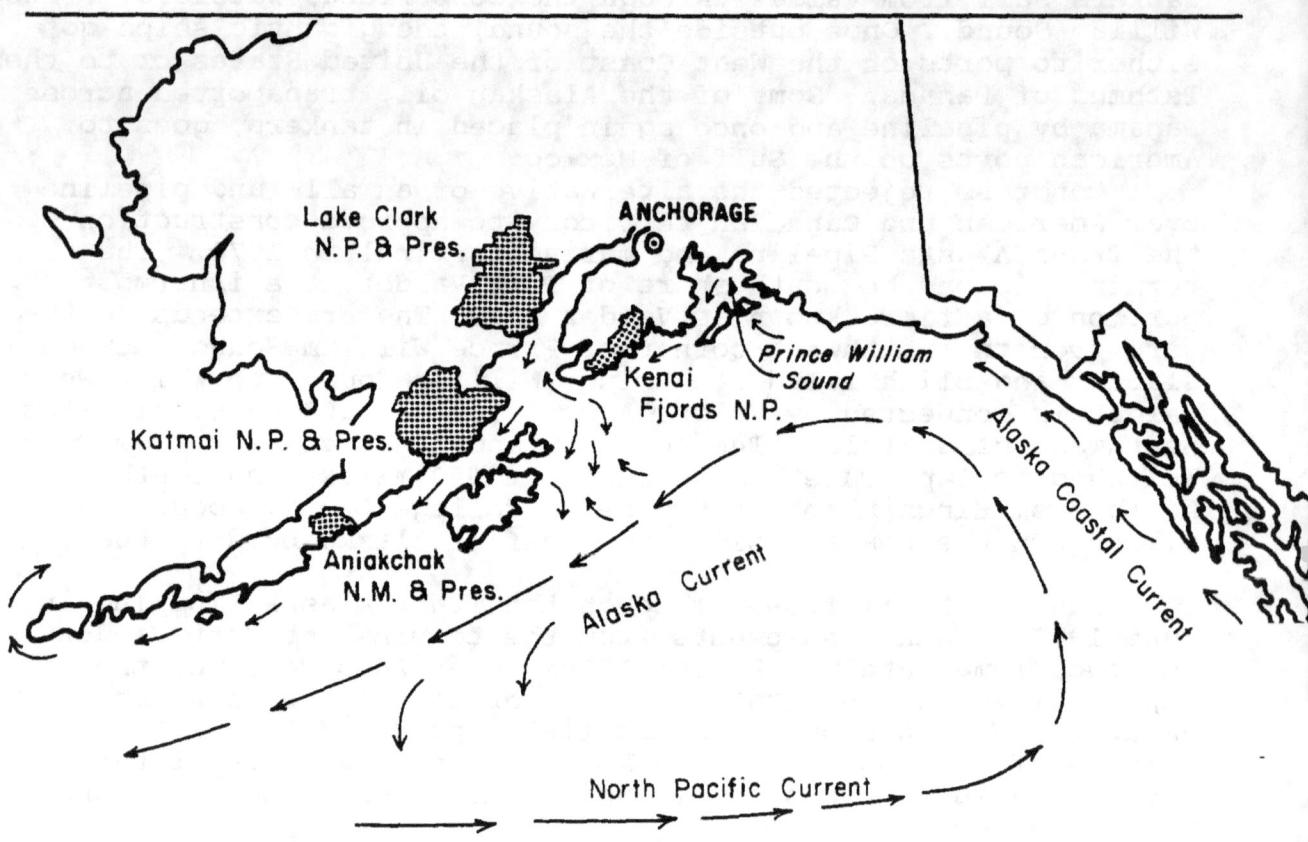

ALASKA OCEAN CURRENTS

Source. University of Alaska Geophysical Institute

$$\frac{953}{80,115}$$

After the <u>Exxon Valdez's</u> oil tanks ruptured on March 24, the spilled oil threatened much of the Alaskan coastline west of Prince William Sound. The Alaska Coastal Current could carry the oil along the coast. Winds from the south could carry it ashore.

"A dragon's breath of swirling death"

The 11 million gallons of North Slope oil released by the ruptured tanks of the Exxon Valdez consisted of highly toxic compounds. One reporter described them as "a dragon's breath of swirling death, the patches of brown oil, tar and sheen."[7]

The oil contaminated the coastline and the food chain, beginning with plankton and continuing through the oiled carcasses of its victims. The swirling death also posed physical and physiological threats to the mammals and birds in its path.

For marine mammals such as sea otters and seals, the principal threat from spilled oil is matting of pelts, later loss of the insulating quality of their fur, and resulting hypothermia. Other potential problems include irritation of eyes and skin, ingestion of poisonous hydrocarbons, and kidney damage. Sea otters are the most susceptible of the marine mammals because they depend upon their fur for insulation. They die of hypothermia and stress when their fur comes into contact with oil. Scavenging land mammals and birds can be poisoned by feeding on oiled carcasses or on food sources such as shellfish found in the intertidal zone. Sea birds that roost on water or forage by diving or surface seizing are particularly vulnerable. Oiled birds die from hypothermia, suffocation, and can contaminate their own eggs with oil, killing the embryo.[8]

Oil contamination is also deadly to plant life. Some spills have killed entire plant communities in the intertidal zone.[9] Observers have also occasionally found damage to plant life such as salt marshes and shoreline vegetation.[10]

Beyond the physical, physiological, and systematic damaged caused by an oil spill, there is aesthetic damage. The oil mars landscapes for years, perhaps centuries. In 1989, Coast Guard and National Oceanic and Atmospheric Administration (NOAA) officials visited the scene of a 1970 Nova Scotia oil spill. They found asphalt three to six feet wide at high tide line and soft tar oozing petroleum in the intertidal zone.[11] A December 1988 spill off the State of Washington affected Olympic National Park. Workers cleaned most of the visible oil and removed more than 750 tons of oily debris. But the park's superintendent, Robert Chandler, testified before Congress "I don't believe we will be able to remove every drop of oil. We will not be able to get the park back to the way it was."[12] Months after the spill, hikers on beaches on Canada's Vancouver Island found tar from the spill oozing out of apparently clean sand.[13]

Resources at Risk

NPS-managed resources placed at risk by the Exxon Valdez oil spill included a variety of marine and terrestrial life in wilderness settings. Much of the area had previously been only lightly touched by human intervention. These parklands were among the most pristine in America, set aside to be preserved unimpaired for all generations.

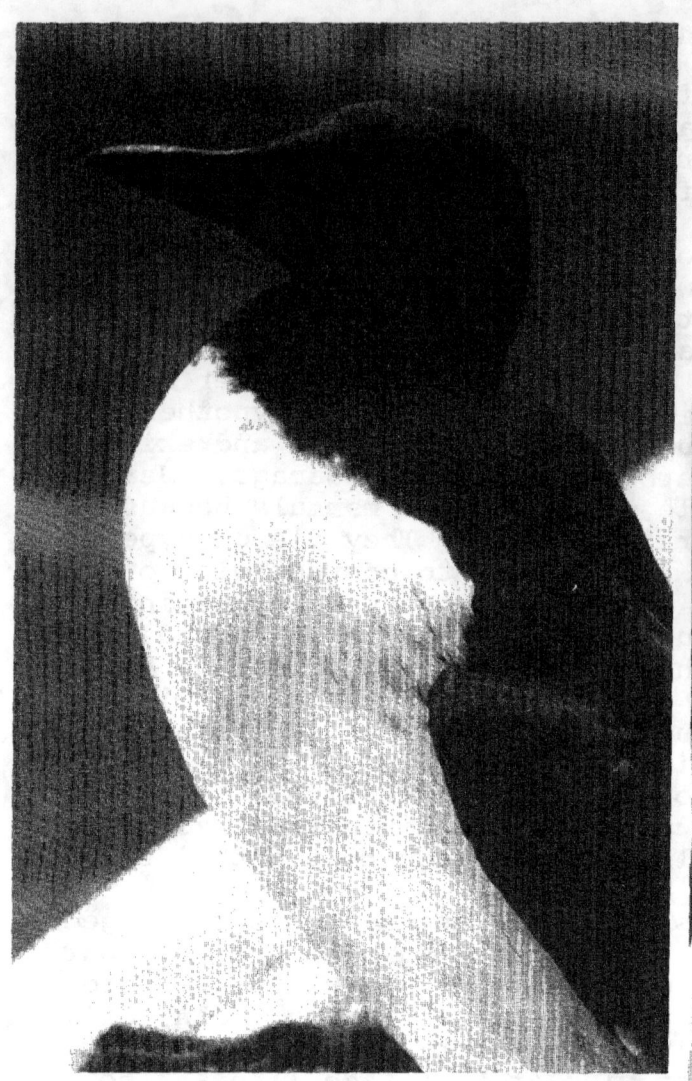

Common Murre. (Photo courtesy of
Karen Jettmar)

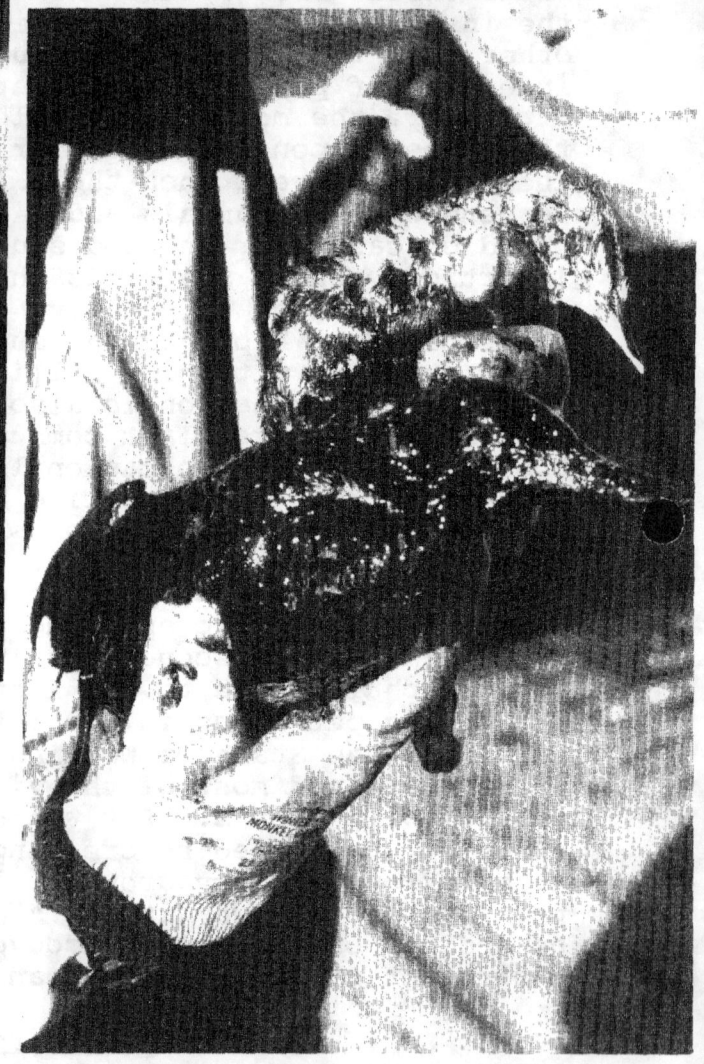

Oiled birds: Common Murre and Red
Necked Grebe. (Photo courtesy of
Karen Jettmar.)

Dead sea otter. (Photo courtesy
of Karen Jettmar.)

(

Dead bald eagle found in tidal
debris on Black Bay beach.
(Photo courtesy of Karen
Jettmar.)

President Carter used executive authority to create Kenai Fjords National Monument in 1978. The monument stretched southwest from Resurrection Bay, a deep fjord about 40 miles west of Prince William Sound, along the Gulf of Alaska coast. Congress transformed the monument into Kenai Fjords National Park in 1980.[14] The park's 395 miles of shoreline are "a priceless necklace of bird rookeries, tidal pools, and water whale playgrounds draped along the Kenai Peninsula's eastern coastline."[15] Twenty-three species of marine mammals including whales, porpoises, dolphins, sea lions, seals, and sea otters inhabit the coastal area. The fjords' cliffs and islands also provide nesting or staging areas for some 250 thousand marine birds of 17 species.[16] This rich coastal environment was home to the "Unizkugmiut" Eskimo at the time of European contact. Archeological sites scattered along this coastline show that prehistoric people with a maritime-based economy, utilized the area as early as 2,000 years ago.[17]

Lake Clark National Park and Preserve lies 100 miles to the northwest of Kenai Fjords, on the eastern shore of the Alaska Peninsula. Established on December 2, 1980, the park covers 2.6 million acres and the preserve covers 1.4 million acres.[18] Its 60 miles of coastline are within the reach of the offshoot of the Alaska Coastal current that circulates through Cook Inlet. Rocky cliffs on the park's coast serve as rookeries for puffins, cormorants, kittiwakes, and other seabirds. Swans and other waterfowl nest on the park's coastal marshes. The park coast serves as an important staging area for migrating wildfowl heading north beginning in April and south beginning in July.[19] Historically, the park's coastal area was used by Tanaina Indians and Eskimos for hunting both land and sea mammals, fishing, and collecting clams. Prehistoric sites of the Alutiiq Eskimo who occupied the area from about 2500 BC to AD 600 and later Tanaina sites are found along the coast.[20]

Katmai National Park and Preserve, first established as a national monument in 1918, was expanded several times over the years. It was further enlarged and given its current designation in 1980 under the Alaska National Interest Lands Conservation Act (ANILCA). On the Alaska Peninsula 160 miles southwest of Kenai Fjords, Katmai's 398 miles of coastline front both on lower Cook Inlet and on Shelikof Strait. It includes islands up to five miles offshore. Like Kenai Fjords, Katmai's shoreline is alive with birds and wildlife. The coast provides habitat for marine mammals and birds, and for moose, bald eagles, and brown bears.[21]

It is also rich in cultural resources, with prehistoric and historic villages, middens and camps spanning the last 6000 years. The region was inhabited by Yupik speaking Eskimo at the time of earliest European contact.[22]

Aniakchak National Monument, established in 1980, lies 120 miles southwest of Katmai on the Alaska Peninsula. Aniakchak's 68 miles of coastline, like those of its companion Park Service units to the east, host a variety of marine and terrestrial life and a cultural history of at least 2000 years.

National Parks in Alaska Impacted by Oil Spill

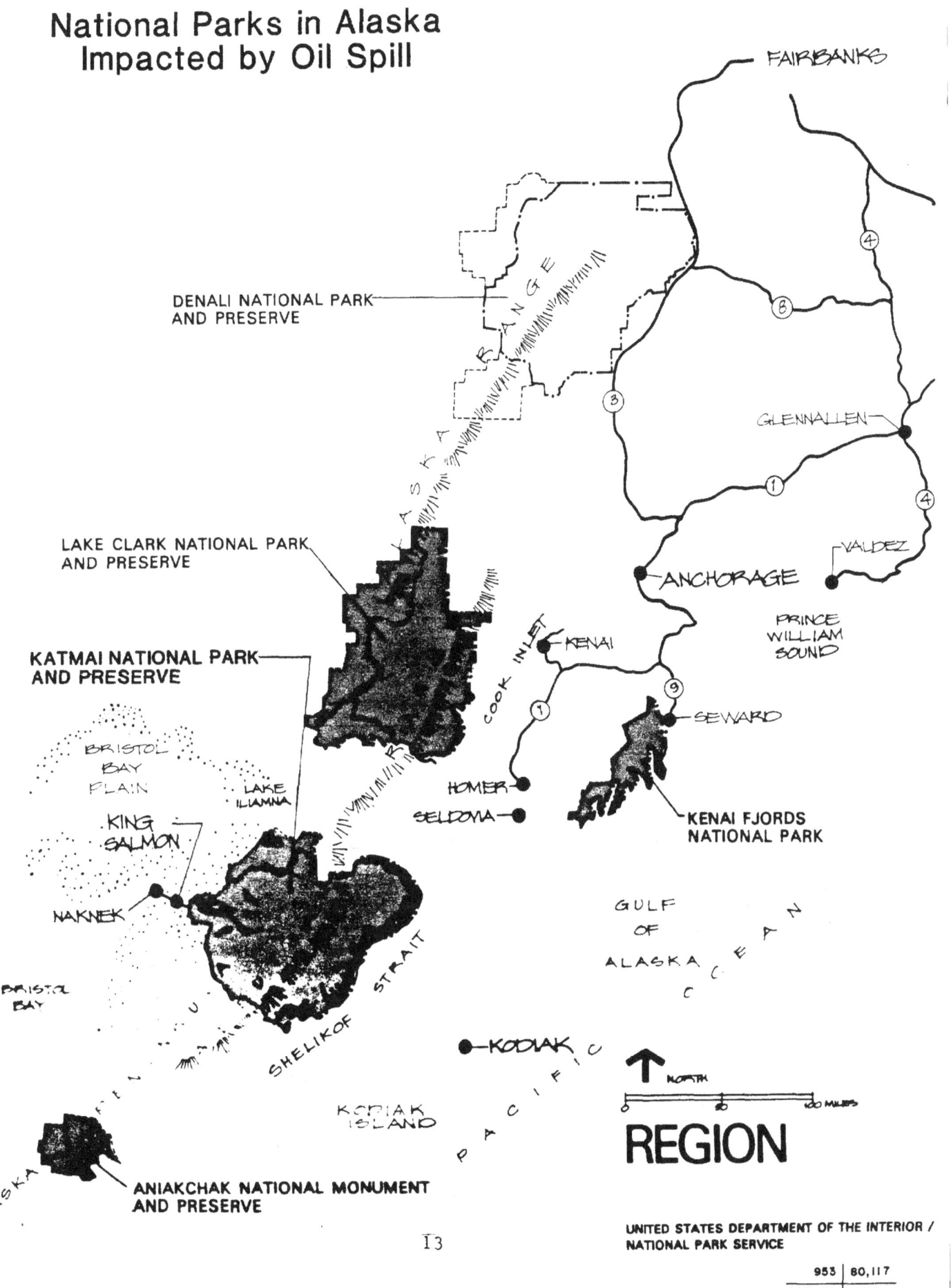

DENALI NATIONAL PARK AND PRESERVE

LAKE CLARK NATIONAL PARK AND PRESERVE

KATMAI NATIONAL PARK AND PRESERVE

BRISTOL BAY PLAIN

LAKE ILIAMNA

KING SALMON

NAKNEK

BRISTOL BAY

SHELIKOF STRAIT

KODIAK ISLAND

ANIAKCHAK NATIONAL MONUMENT AND PRESERVE

ALASKA RANGE

FAIRBANKS

GLENNALLEN

VALDEZ

ANCHORAGE

PRINCE WILLIAM SOUND

KENAI

COOK INLET

HOMER

SELDOVIA

SEWARD

KENAI FJORDS NATIONAL PARK

GULF OF ALASKA OCEAN

KODIAK

PACIFIC

NORTH

0   50   100 MILES

REGION

UNITED STATES DEPARTMENT OF THE INTERIOR / NATIONAL PARK SERVICE

953	80,117
FEB 90	ARC

The four park units threatened have between them about 920 miles of coastline and 9,400,000 acres in aggregate. Each park had a staff of six to eight in 1989. Katmai National Park and Preserve and Aniakchak, a monument, shared a staff of eight. Aniakchak had no dedicated staff. Only Kenai Fjords had a boat, a limited capability 30-footer. The Alaska Region, which oversees these and 10 other National Park System units, had a total staff of only 308.[23]

This minimum staffing influenced how the Park Service in Alaska responded to the emergency created by the Exxon Valdez oil spill. An absence of scientific documentation on each park unit's coastline compounded the problem created by minimum staffing. Remote and seldom visited in the past, the coastlines had many unknowns about their cultural, faunal, floral, and recreational resources. The NPS Alaska Regional Office recognized this lack of knowledge and had proposed an eight-million dollar initiative to correct it, but nationwide funding shortages precluded its acceptance.[24] When oil-imposed injuries to its coastlines seemed likely, the NPS had to quickly do pre-oiling assessments to document the condition of coastal resources.

National and Regional Response Mechanisms

National Response System

Initial response to the oil spill on March 24 fell most heavily not on the Park Service, but on three other Department of the Interior (DOI) elements. These were the Regional Environmental Office in Anchorage, the Alaska Region of the Fish and Wildlife Service, and the Alaska Fire Service, a unit of the U.S. Bureau of Land Management. The Bureau of Land Management's Pipeline Monitoring Office contacted Paul D. Gates, Regional Environmental Officer for DOI, at 0115 on March 24 with news of the oil spill.[25]

Gates, as the Department of the Interior member of the federal government's Regional Response Team for Alaska, had dealt with several oil spill incidents. The Regional Response Team (RRT) is part of the National Response System. The Alaska RRT coordinates Federal activity and advises the Federal On-Scene Coordinator (OSC). The OSC is the Federal representative with action authority. For inland incidents, the Federal OSC is an Environmental Protection Agency representative; for on-water incidents the Federal OSC is a U. S. Coast Guard (USCG) representative.

The National Response System resulted from concern over the nation's ability to handle oil spills of the magnitude of the Torrey Canyon incident. In that incident in 1967, a tanker spilled over 26 million gallons of oil off the coast of England. By 1989, the National Response System included 14 federal agencies. The agencies participate on the National Response Team (NRT). Most also have representatives on teams for each of the ten federal regions in the contiguous 48 states and for Alaska, the Caribbean, and the Pacific Basin. The teams coordinate response to oil discharges and hazardous substance releases. The NRT operates under the authority of the National Contingency Plan, promulgated as a Federal regulation in 1973 under the Clean Water Act (CWA) and the Comprehensive Environmental Response, Compensation, and Liability Act (CERCLA).[26]

Gates served as DOI's representative to the Regional Response Team for the Exxon Valdez incident. He also functioned as a conduit for information passed to and from DOI headquarters in Washington and Interior field offices in Alaska. Besides these activities, Gates established and supervised a DOI Coordination Center in Anchorage. In each of these roles, Gates had continual involvement with the National Park Service.

Gates contacted Pamela A. Bergmann, the Regional Environmental Assistant as soon as he was notified of the spill. Bergmann had just returned from serving as the DOI on-scene representative at an oil spill in Dutch Harbor. On March 24, almost before she had unpacked from her trip to Dutch Harbor, Gates sent Bergmann to set up a DOI Coordination Center in Valdez.

DOI bureaus sent representatives to Valdez to assist Bergmann. Bergmann asked that Page Spencer, an ecologist with the NPS Alaska Regional Office, come to Valdez. Spencer was not available at the time, and although the Alaska Regional Office offered to send William B. "Bill" Lawrence, Chief of Environmental Compliance Division, he was never sent. The fact that a NPS representative was not present in Valdez to support the DOI Operations Center and represent NPS interests was regretted later. Although there was continuous communication between Valdez and the Regional Office in Anchorage, a liaison at the Command Center in Valdez would have proved useful when DOI-NPS misunderstandings arose.[27]

Bill Lawrence was a member of Gates' team of DOI representatives. On March 29, Lawrence was notified by Gates that the Coast Guard predicted that oil would leave Prince William Sound, putting NPS areas at risk. From that point on Lawrence participated in RRT meetings and worked almost full-time on the oil spill. Among Lawrence's contributions was an extensive background in emergency response and experience with oil spills. Lawrence served as a liaison between the Regional Response Team and the National Park Service.[28]

One of the responsibilities of the NPS Environmental Compliance Division is oil spill planning and prevention. The possibility of an oil spill had been anticipated, and at the time of the Exxon Valdez incident a regional plan had been developed. An oil spill contingency plan for Kenai Fjords National Park had just been completed and was under review, and the framework for assisting the other parks in developing individual plans was in place.

DOI Agencies

The Fish and Wildlife Service, with its broad responsibilities, was the DOI bureau most affected by the spill during the first five days. Its personnel inventoried migratory birds and sea otters in Prince William Sound and monitored bird and otter rescue and rehabilitation efforts.[29]

The Alaska Fire Service became involved with the spill when Exxon requested that the Type-I Alaska Incident Command Team (ICT) come to Valdez. The Type-I Alaska ICT and its 17 counterparts scattered throughout other areas of the United States are part of a National Incident Management System. The teams train and serve principally to manage response to wild fires, but they are considered "all-risk," and also participate in other types of incidents such as disaster relief.[30] The Type-I teams have the system's most experienced personnel and are used to deal with complex and large incidents. There are also Type-II and III teams, used to deal with lesser incidents or with the later stages of incidents that initially required Type-I teams.

Alaska Incident Command Team

The Type-I Alaska ICT became involved in the Exxon Valdez oil spill when Exxon requested its help. Don Cornett, Exxon's Project Manager at Valdez, made the request in a March 27 telephone conversation with Les Rosenkrance, the U.S. Bureau of Land Management's Associate State Director for Alaska.[31] The Coast Guard too asked about the ICT. The Coast Guard is the federal government's On-Scene Coordinator for on-water oil spills. Coast Guard officials at Valdez asked if the ICT could set up five onshore camps for cleanup crews that would be working in Prince William Sound.[32]

It is not clear if the agencies at Valdez asked the Alaska ICT to come to Valdez only to set up the camps. Some might have thought that the ICT could have a broader role in management of the oil spill response.[33]

Whatever the motivation, the Alaska ICT mobilized and went to Valdez on March 28. Its intense and wiry commander, Dave Liebersbach, took with him the habit of command gained in 20 years of fire fighting and smoke jumping, and the team's core staff. The core staff consisted of Don Wahl, Safety Officer; Dixie Dies, Information Officer; Marv Robertson, Planning Section Chief; Tom Goheen, Operations Sections Chief; Don Fuller, Logistics Section Chief; and Ron Knowles, Finance Section Chief.[34]

This staffing, which could and later did expand to meet needs as they arose, provided the framework for five critical functions in Incident Command operations. These functions are Command, Operations, Logistics, Plans, and Finance. Command provided general oversight and direction. Commander and command staff oversaw safety, information, and interagency coordination. Operations accomplished planned activities. Logistics provided

16

services and supplies needed to support operations. Plans tracked equipment and personnel working on the incident and provided information about conditions to the Incident Commander so that decisions could be made. Finance tracked all expenditures and assured accountability for personnel time.[35]

The core team members filtered into Valdez throughout the day of March 28. Liebersbach met with Bergmann and Coast Guard and Exxon officials. Late in the day Dies, Goheen, and Wahl met at the Valdez Coast Guard Station for a briefing.

By the morning of March 29, the core team members had concluded that no one in Valdez wanted the services of the ICT. The team remained in Valdez throughout March 29 without being drawn into the activities of any of the agencies there that were responding to the oil spill.[36] Exxon's decision to house cleanup workers on barges and boats eliminated the need for an organization to set up on-shore camps.

The organizations that might have taken advantage of the ICT's expertise for other activities had, by March 28, already put different management mechanisms in place. Chugach National Forest was the principal federal land manager initially affected by the spill. The Forest Service chose to take a low-key approach and worked directly with Exxon, although later it turned to the Incident Command System for help. The U.S. Fish and Wildlife Service also responded cautiously to the spill. It became more active only as the spill directly affected wildlife refuges.[37] Other potential government users in Valdez of the ICT were the State of Alaska and the U.S. Coast Guard. Each activated its own in-place system for dealing with emergencies. Exxon, the potential private user of the ICT, chose simply to supplement its day-to-day operations to deal with the spill. As a result, officials released the team from Valdez on the evening of March 29. It was called upon immediately to help the National Park Service respond to the oil spill.[38]

Response Resources at Kenai Fjords

Kenai Fjords National Park, headquartered at Seward 123 miles south of Anchorage on Alaska Highway No. 1, was the first park to recognize an immediate threat from the Exxon Valdez oil. Minimally staffed, the park was fortunate in the preparation of assigned personnel to deal with the problem it faced.

Superintendent Anne Castellina came to the park with extensive experience in working with groups of people. Her prior assignments included field interpretation roles and training activities at the Park Service's Harpers Ferry Center. Chief Ranger Peter Fitzmaurice was well-acquainted with the Incident Command System and seasoned by several summers' work in dealing with ICTs managing wildfire responses at other parks. Resource Specialist William D. "Bud" Rice, was intimately familiar with the park's coastline and had just completed a master's project on glaciers and climate that involved offshore currents.[39]

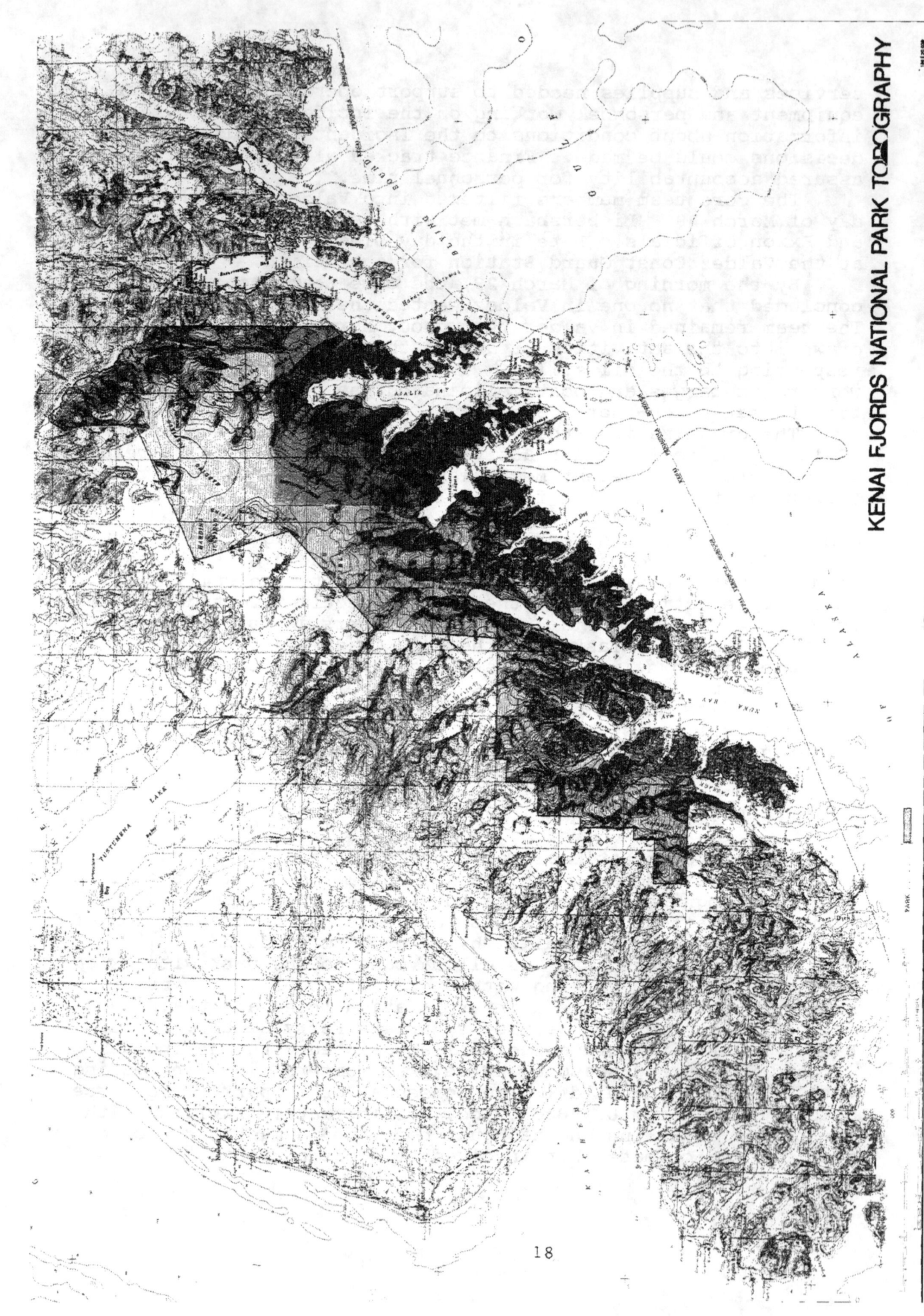

KENAI FJORDS NATIONAL PARK TOPOGRAPHY

18

Coincidentally, Rice had just completed and Fitzmaurice was in the process of reviewing an oil spill contingency plan for Kenai Fjords National Park.

Identification of the Threat

When Rice, on annual leave at the time of the spill, returned on Monday, March 27, Castellina asked for his opinion about the possibility of oil coming out of Prince William Sound and into the waters off Kenai Fjords. By Wednesday, March 29, Castellina, Fitzmaurice, and Rice were concerned enough to discuss preparations with contacts in Seward. Fitzmaurice, as oil spill coordinator for the park, consulted the park's just drafted oil spill contingency plan. Late in the day, at 1600, Castellina and Fitzmaurice attended an emergency meeting with municipal officials at the Seward City Hall.[40]

Decision to Call for Help

Castellina also called David B. Ames, Associate Director for Operations, and Acting Regional Director, at the Alaska Regional Office to ask for assistance. Ames, a Park Service veteran, came to his Alaskan job from the superintendency at Hawaii Volcanoes National Park. Eruptions, fire, and law enforcement situations kept that park in emergency conditions about one-third of the time. That experience left Ames with no hesitancy in dispatching or calling for help when needed, or in reacting decisively in emergencies. When Castellina requested $10,000 to hire an additional ranger, Ames approved immediately. The ranger was to be stationed at Nuka Bay, almost 80 miles southwest of Seward on the outer coast of the Kenai Peninsula and near the western boundary of Kenai Fjords National Park. The new staffer was to monitor any impact from the oil spill.[41]

After Castellina's call, Richard G. "Rich" O'Guin, Chief of the Protection and Ranger Activities Division at the regional office, alerted Gates. The Regional Environmental Officer then called a meeting of the Regional Response Team for 1400. At this meeting, Interior officials were reluctant to initiate action because the Coast Guard was the designated federal lead agency for oil spills.

While the Regional Response Team was meeting, Ames told O'Guin that he had decided to call in an ICT to help the NPS response to the oil spill. Ames' staff endorsed the decision. Steve Shackleton, his Law Enforcement Specialist, pointed out that the spill was a classic case for ICT use.[42]

The Interagency Fire Center in Boise, Idaho is the agency through which requests for National Incident Command System assistance are made. O'Guin coordinated with the Alaska Fire Service in Fairbanks, the Interagency Fire Center in Boise, and the NPS Ranger Activities Division at Park Service headquarters in Washington, D.C. As a result, fire officials reassigned the ICT leaving Valdez to the Park Service.[43]

O'Guin called Castellina to tell her the results of the Regional Response Team meeting. He also advised her of the availability and capability of the ICT expected to be demobilized from Valdez.[44] O'Guin's information supplemented a conversation between Castellina and Ames about ICTs while both were attending a 1988 class at the Federal Law Enforcement Training Center in Glenco, Georgia. Despite concerns about adding another layer of overhead to park operation, Castellina concurred with the idea of dispatching an ICT to Kenai Fjords.[45]

Washington Reaction to ICT Help

Organization

Within the Department of the Interior's Washington office, the Office of Environmental Project Review (OEPR) took responsibility for organizing oil spill response for the department. At the same time, Vern Wiggins, Deputy Undersecretary for Alaskan Affairs, assumed leadership of an ad hoc group formed to oversee oil spill matters. Wiggins later chaired daily briefings at which Interior bureaus reported on oil spill activity of particular interest to them. Concurrently, Denis P. Galvin, Deputy Director of the NPS, became the Service's liaison with Interior for oil spill matters. Galvin also became the Alaska Region's Washington contact for oil spill matters.[46] While Wiggins and OEPR addressed broad departmental concerns, including NPS issues, Galvin focused on Park Service matters.

Perspectives

The differing perspectives were soon clear. A series of environmental laws largely shaped departmental concerns. These laws were the Clean Water Act (Federal Water Pollution Control Act, 33 U.S.C.A. Sections 1351-1387), CERCLA (Comprehensive Environmental Response, Compensation, and Liability Act, 42 U.S.C.A. Sections 9601-9675), and several other statutes and regulations that provide the context for dealing with a major oil spill. Many of these laws address the question of financial responsibility but are unclear. As a result, departmental officials directed their attention to the question of who would pay for oil spill response activity.[47]

Galvin had a background of technical and managerial experience in many National Parks and NPS offices. Most recently, as Deputy Director, he had dealt with the Yellowstone fires of 1988. Galvin continued to serve as Deputy Director until April 18 when the Bush administration put a new top management team in place for the Park Service.

Galvin took the position that the oil spill was an emergency. Section 101 of the Service's budget act authorizes expenditure of funds from any source to deal with an emergency. Most of his initial discussions with departmental officials regarding the oil spill focused on funding. The questions put to him were:

Are you prepared to spend Park Service money, and do you have the authority to spend Park Service money for oil spill response? Galvin said yes. If all else failed, the Park Service could freeze funding for the Natural Resources Preservation Program and use that money to pay for oil spill work. The oil spill work was inventorying and monitoring existing conditions, activities for which Congress appropriated money to the Natural Resources Preservation Program.[48]

The Exxon Valdez oil spill occurred in the midst of a national debate over whether to open the Arctic National Wildlife Refuge (ANWR) to oil exploration. The debate lined up conservationists on one side and pro-development industry and government officials on the other side. President Reagan and then President Bush came out for ANWR exploration. The administration's initial low-key response to the oil spill generated press speculation about the relationship between the push to open ANWR and the spill. Some said that the quiet response was an attempt to downplay negative associations between the spill and oil development in ANWR.[49]

Interior Department officials' close scrutiny of Park Service response to the spill gave some the impression that the officials wanted as little furor as possible about the spill and its environmental impact. Constant questioning about the funding and appropriateness of Park Service response activity contributed to that impression.

Questions on money and expenditures came up in almost every meeting on the oil spill that Galvin attended. Other departmental concerns were expressed in a March 31 meeting. Those present included Lou Gallegos, designate Assistant Secretary for Policy, Budget, and Administration; Deputy Undersecretary Wiggins; Mary Anne Bach, Deputy Assistant Secretary for Fish, Wildlife, and Parks; representatives of the Department of the Interior's Solicitor's Office, and representatives of OEPR. At this meeting, departmental officials asked Galvin "What was the Park Service doing? Were we coordinating with everybody else? Why had we called up the Incident Command Team?" In this meeting, and later, there was concern that the Park Service might be acting precipitously. There was considerable skepticism and some criticism of the Park Service for reacting hastily to a threat that some observers did not think would materialize.[50]

Boyd Evison, Alaska Regional Director for the NPS, also encountered this skepticism and resulting caution from Washington officials. Evison, a 29-year Park Service professional, came to his Alaska position after several superintendencies and appointments on bureau and department staffs in Washington. This service had given him some knowledge of the Incident Command System and a sense of ease in dealing with high-level executive branch officials and with members of Congress. When the Exxon Valdez ran aground, he was on his way to Washington to attend a Senate appropriations hearing.[51]

When he reached Washington, Evison talked by telephone with Ames, to whom he had delegated Acting Regional Director responsibilities during his absence from Alaska. From Ames, Evison learned about the decision to call in an ICT and of Wiggin's disapproval of that decision. Evison backed Ames. He knew that neither Kenai Fjords or Katmai had the staff to handle the emergency they might face.[52] Despite Wiggins' opposition, Evison endorsed Ames' decision to call in the ICT. Evison advised Ames to instruct the ICT to focus particular attention on three areas: gathering pre-impact data on resources at risk, arranging to track oil movement, and arranging for post-impact monitoring.

At the appropriations hearing, senators asked Evison about the spill. He explained the opportunity to call in the ICT and the need to inventory the park coastlines before the oil struck.[53]

Evison also clashed with Wiggins at a briefing for the Secretary of the Interior. At the briefing, Wiggins assured the Secretary that oil wasn't going to get out of Prince William Sound; and, if it did, it would just be tiny balls of inert stuff. Evison uncomfortably advised the briefing group that the oil was already at the entrance to Resurrection Bay. It would almost certainly strike the coast of Kenai Fjords National Park and probably coastlines of other parks to the southwest.[54]

Returning to Alaska over the weekend of April 7, Evison was able to see for himself the oil hovering off the park's coastline. He also watched the ICT in action. Then he went back to Washington to testify at hearings in the House of Representatives held by Congressman Bruce F. Vento (D-MN).[55]

The regional director's Congressional contacts plus press reports and pressure from Alaska's state officials and senior Senator soon overwhelmed departmental criticism of Park Service response to the oil spill. After his hearings, Congressman Vento was unrestrained in his comments. He said that what national press called tepid, understaffed response by the Interior Department, "represented a broken promise to the American people." According to the article that quoted Vento:

> Park Service employees were excluded from the Interior Department's top-level spill assessment meetings after one employee challenged an early, upbeat report on the damage. . . .Top Interior Department officials are downplaying the spill's damage to Alaskan parks so as not to interfere with administration efforts to promote oil exploration in the fragile Alaskan wilderness, according to Interior Department sources.[56]

Support for Park Service actions at the national level was complimented by local endorsement from an unusual source. The Anchorage Times, very pro-development and usually critical of the NPS, editorialized that the Service acting:

on the theory that moves to protect the park and assemble the mechanism [the ICT] to provide an organized response were the only responsible courses of action...the action by the Park Service is commendable.[57]

This flurry of local and national support came on the heels of a major political breakthrough. Alaska's Sen. Ted Stevens had returned the previous week from a visit to Prince William Sound and the communities of Cordova, Valdez, Seward, Homer, and Kodiak. Deeply affected by the oil-wrought devastation in Prince William Sound and sensitive to the worries of his constituents in communities to the southwest of the Sound, Stevens arranged for Alaska's Congressional delegation (himself, Sen. Frank Murkowski, and Rep. Don Young, all Republicans) to meet with the President.[58]

Immediately after the meeting, President Bush announced on April 6 that he was appointing Secretary of Transportation Skinner his personal liaison for the spill. Admiral Paul A. Yost, Commandant of the Coast Guard and Coast Guard Vice Admiral Clyde E. Robbins were to go to Alaska and take personal charge of the spill response. The President also ordered the Department of Defenses to make its facilities, equipment, and personnel available for oil spill response as needed.[59] The outpouring of public support, media encouragement, and awakened concern at the Presidential level about the oil spill seemed to calm further departmental alarms that the Park Service was overresponding to the calamitous spill.

In retrospect, the early departmental worries are under- standable. Wiggins and most of his departmental colleagues were recent appointees of the Reagan administration continued in the Bush administration. Bush officials were new to Washington and unfamiliar with the Park Service bureaucracy. All knew that President Bush was in favor of ANWR development and they did not want to obstruct it. At the same time, the Washington officials were receiving conflicting information from different sources. On the one hand, one DOI bureau with agents on-scene, the Fish and Wildlife Service, did not think the oil spill did or would endanger National Parks. NOAA experts at this time were pre- dicting that the oil would not reach the National Parks to the west of Prince William Sound. On the other hand, another DOI bureau also with agents on-scene, the NPS, thought the oil spill endangered National Parks and dealt with the situation as an emergency. The resulting hesitation often translated at the field level into a belief that departmental officials did not support Park Service efforts to deal with the oil spill.[60] The combination of political interpretation, transition uncer- tainties, and conflicting information naturally enough led to hesitation about the Park Service's aggressive response to oil spill dangers.

The doubts should have, but didn't end. On April 20, when Evison attended a briefing for Adm. Yost at Elmendorf Air Force Base, Coast Guard and NOAA officials denied that oil had hit

23

Katmai. The Park Service by then "had a jar full...of samples" of oil from its beaches. At the same meeting, Walt Stieglitz, Alaska Regional Director for the U.S. Fish and Wildlife Service, sat next to Evison. He remarked that the Park Service was over-reacting to the oil threat.[61] This attitude continued to affect Park Service relations with its sister agencies and departmental officials throughout the incident.

The difference seemed to lie in the purposes for which the agencies were responding. Galvin, Evison, and other Park Service officials intended to meet the Service's responsibilities under the 1916 organic act that established the NPS. Departmental officials, other Interior bureaus, and non-Interior agencies were operating within the limitations imposed by the Clean Water Act and CERCLA. These provide for recovery of costs incurred by the federal government to restore or replace natural resources lost as a result of an oil spill. Marching with different orders, the Park Service, although doing its duty, sometimes seemed out-of-step. Calling in the ICT was the first sign.

Decision to Call in the Alaska Incident Command Team

Kenai Fjords National Park

Kenai Fjords National Park was the first Park Service unit to use the ICT to deal with the Exxon Valdez oil spill. Local governments, state and federal agencies, and finally Exxon Corporation itself eventually came to rely on the ICT for assistance in dealing with the emergency. Although the National Incident Management System had much experience in dealing with fire suppression, law enforcement situations, and search and rescue efforts, the Exxon Valdez oil spill was the first time an ICT provided assistance in this kind of emergency. That unusual application of the team became even more unusual because of the variety of entities that participated with the team.

The Alaska ICT cleared Valdez at 2300 on March 29. By 0800 it was in Anchorage to meet with NPS regional staff before going on to Seward. At 1100 on March 30, the ICT met with NPS Acting Regional Director Ames, Bill Lawrence, and others on the regional staff. Ames charged the ICT with coordinating efforts to protect the Kenai Fjords coastline and wildlife from the approaching oil.[62]

Even before the ICT arrived in Seward, Castellina had delegated to Liebersbach the authority to act for the park in handling the oil spill.[63] By 1730 on March 30, the ICT was in Seward and meeting with Superintendent Castellina. Thirty minutes later, Dies and two City of Seward officials, Chris Gates and John Gage participated in a public meeting on the oil spill. Gates was the city's Director of Maritime Operations. Gage was the city's Fire Chief and Director of Emergency Operations.

Preparations

Castellina arranged for the ICT to be billeted at the Army's Seward Recreation Camp on the outskirts of Seward. Fitzmaurice arranged for the ICT to use an unoccupied U.S. Forest Service seasonal housing unit as an operations center. Rice began to prepare habitat maps and resource-at-risk maps. Other park staff made photocopies of Kenai Fjord's oil spill plan to be distributed to the ICT. All turned to Fitzmaurice for advice based on his prior experience in working with incident command teams on fires.[64]

As some of the Kenai Fjords staff prepared for arrival of the ICT on March 30, other staff members took their first look at the spilled oil. Rice and Spencer took annual leave and flew over the Exxon Valdez and the oil slick. Seeing "black waves washing up on Knight Island and the lobe of the oil at the south tip of Montague Strait" confirmed Rice's fears that Kenai Fjords National Park would get oil -- "Lots of it."[65]

The overflight ended at the Girdwood Airport, about 100 road miles north of Seward. As they drove back towards Seward, Spencer and Rice discussed what needed to be done at Kenai Fjords to prepare for the oil spill. The result was a hastily drafted plan that became the basis for Kenai Fjords' response to the oil spill. In the rough plan, Spencer and Rice identified the need for specialists to document the condition of the park before the oil arrived. They also projected a requirement to assess the damage after the oil had washed ashore.[66]

The plan was drafted in terms of what resources would first be struck by the oil. These included elements such as water quality in the water column, fish, plankton, and intertidal organisms such as crustaceans and sea mammals. The timing of the spill was just right to have the worst impact. Sea birds were returning to the coast. Whales were just beginning to migrate along the coast. Carrion feeders such as bears and eagles would be on the beaches. Contamination of energy, food, and life through time and space was a critical issue.[67]

After Rice returned to Seward, city officials asked him for a list of the ten most significant salmon streams in the Resurrection Bay area.[68] Rice drew on his own expertise and information contained in maps and data sets produced by the Cook Inlet Aquaculture Association which were compiled from surveys conducted by the Alaska Department of Fish and Game. Tom Schroeder, the state fisheries biologist responsible for the area was stationed at Homer, a fishing port 80 miles west of Seward. He later confirmed and approved the first draft of the priority list. The result was a list of significant salmon streams in the Resurrection Bay/Gulf Coast area near Seward that he believed to be protection priorities. The list, in descending order of priority, included Resurrection River, Desire Lake, Delight Lake, Tonsina Creek, Pederson Lagoon, James Lagoon, Thumb Cove, Humpy Cove, Two Arm Bay, and Quicksand Cove. The priority list

Ineffective boom.   (Photo courtesy of Karen Jettmar.)

Assembling boom in Seward.   (Photo courtesy of Karen
Jettmar.)

included estimates of the amount of boom need to protect these streams, reaching a total of 5,500 feet.[69]

"Boom" refers to barriers designed to keep oil from floating into particular areas or contained within particular areas. In addition to these "curtain" booms, "deflection" booms are sometimes used to guide oil away from particular areas. The boom usually consists of plastic coated foam board with lead ballast attached to the bottom. A lighter version of boom is made of thin plastic film with plastic floats attached. Those familiar with the subject often describe boom in terms of the depth of the boom from top to bottom, for example 24-inch or 86-inch boom. Thirty-six-inch boom, used frequently, extends 12 inches above and 24 inches below the water's surface. Boom is usually tied at either end to metal fence posts or trees on shore, although it is sometimes affixed to sea anchors.[70]

In general, boom is not effective when stretched over long distances or when it is subject to currents, strong tides, or winds. Even when boom is effective, it must be constantly maintained. Material a boom does "bar" must be removed from in front of the boom or the boom will become "entrained." Then the boom allows the material to pass over or under it.

The next day Rice revised his estimate. He still listed Resurrection River (2000-4000 feet of boom); Tonsina Creek (1000 feet of boom); Thumb Cove (500-1000 feet of boom); Humpy Cove (500 feet of boom); Pederson Lagoon (1000 feet of boom); Delight Lake (500 feet of boom); James Lagoon (500 feet of boom); Desire Lake (1000 feet of boom); Two Arm Bay (1000 feet of boom); and Quicksand Cove (1000 feet of boom). Boom required totalled 11,500 feet.[71]

The revised priorities resulted from concern for local priorities, which all agreed warranted special attention since the City of Seward owned initial supplies of boom being used. This ready cooperation and sharing of resources forecast a remarkable joint venture that would guide Park Service oil spill operations over the next few weeks.

Comment

Up to this point, NPS enjoyed a combination of fortunate coincidences. These allowed it some advance preparation in dealing with the oil spill. Unlike the Forest Service, which suddenly found lands it managed awash in deadly petroleum product, NPS had some warning. An ICT was available; Ames knew how to use it; and Castellina, Fitzmaurice, and Rice were alert to the dangers threatening their park. Galvin's support at the highest levels of the National Park Service during the first few weeks following the spill, and Evison's work with Congress, as well as NPS and Interior officials in Washington and Alaska, were valuable. Even so, NPS faced difficulties in convincing others in Alaska and in Washington that the oil threatened park coastlines. Such difficulties complicated subsequent aspects of NPS response to the spill. These included the critical one of

establishing command, control, and communication for necessary
staff and field operations in the pre-oiling phase of spill
response.

Journalist dripping with oil on Seal
Island beach. (Photo courtesy of Karen
Jettmar)

CHAPTER 2 - COMMAND, CONTROL, AND COORDINATION

- Overview
- Chain-of-Command
- Line Officer's Briefing
- Multi-Agency Coordinating Group
- Coast Guard Coordination
- Expanded Scope of Operations
- Comment

Overview

Command, control, and coordination includes determining what tasks needed to be done, how to do them, where to get the resources to do them, and directing their accomplishment. These were major challenges for those responding to the Exxon Valdez oil spill. The oil itself was elusive, difficult to see either at water level or from the air. The geographic distances to be covered were vast and subject to violent weather changes. The response mechanisms were complex and that complexity was further complicated by the number of interested agencies.

While the Incident Command Teams provided logistical and operations support in the field, overall management of the Park Service response to the spill continued from the regional office in Anchorage. No precedent, and few policies and procedures, existed for responding to an oil spill that would extend over hundreds of miles, several months and involve complex inter-relationships with federal, state and local governments.

The adaptations of existing structures to accommodate the factors outlined above are addressed in the following narrative. These include chain-of-command relationships in which ICS and NPS management structures were integrated, inclusion of a Multi-Agency Coordinating (MAC) Group in that integration to provide for the interests of the multiple interested agencies, and relationship of the ICS-NPS-MAC Group apparatus to the Coast Guard.

Chain-of-Command

In the ICS, the chain-of-command for response operations normally runs from land manager to the ICT commander to his or her staff. The commander's staff in turn oversees field operations. Circumstances and statutory requirements made lines of authority more complex for the NPS response to the Exxon Valdez Oil Spill. The first land manager involved, the Superintendent of Kenai Fjords National Park, established initial guidelines for ICT operations. But the number of agencies involved soon led to creation of a MAC Group. Coast Guard authority as Federal On-Scene Coordinator (FOSC) for on-water oil spills also added to the complexity of the situation. As the spill response effort evolved, the ICT at Seward established outlying branches. Thus the story of command, control and coordination must deal with the land manager's original briefing (Line Officer's briefing in ICT parlance) to the ICT, the formation and activity of the MAC Group, and branch operations of the ICT.

Line Officer's Briefing

When key people of the ICT were in place in Seward on March 31, Anne Castellina briefed them on what they faced. Oil from the <u>Exxon Valdez</u> spill was moving out of Prince William Sound through Montague Strait. The Alaska Coastal Current would carry it past the headlands and beaches of Kenai Fjords National Park. Forecasters expected currents to push the oil onto windward sides of fjords in the park and onto offshore islands. Certain wind conditions could carry the oil into Resurrection Bay and deep into the park's fjords. In addition to park lands, shoreline managed or owned by the City of Seward, private owners, the State of Alaska, and the U.S. Fish and Wildlife Service was at risk. Potential participants in a unified command of ICT operations included the Alaska Air National Guard, the City of Seward, and the U.S. Coast Guard.[70]

Park staff was limited, but available to help the ICT. Besides Castellina, Peter Fitzmaurice, and Bud Rice, the Kenai Fjords staff included Karen Gustin, Chief of Interpretation; Michael Tetreau, Plant Biologist; Bill Stevens, Maintenance Worker and Boat Operator; Diana Thomas, Interpretive Specialist; Lola Cabaniss, Administrative Technician; and Ida Murdock, Administrative Assistant. Castellina assumed the role of Land Manager's Representative to the ICT and assigned Rice to act as Resource Advisor to the ICT. She delegated authority to Dave Liebersbach to manage Kenai Fjords National Park's response to the oil spill.[71]

Castellina defined the park's priorities for the ICT. In order of priority, resources at risk and of special concern included:

1. salmon streams and salmon fry
2. bird congregation areas
3. seal haul-out areas
4. beach areas containing fragile or endangered plant species
5. areas of significant bivalve concentrations
6. birds and mammals which feed on other dead land animals
7. areas of particular scenic value (the entire coastline)

The briefing identified bears coming out of hibernation and the oil itself as special hazards to be watched for in field operations. In concluding, Castellina alerted the ICT:

30

Politically, this is a highly charged crisis. Our mandate from Interior stresses coordination through them, a virtual lockup of anything to do with the press, and coordination of any actual mitigation efforts only through the Coast Guard. At this point our goal is to collect resource data only.[72]

Multi-Agency Coordinating Group

Forming the Group

Shortly after arriving in Seward, Liebersbach advised Castellina to establish a MAC Group. Provided for in the ICS, such a group could coordinate efforts of various agencies with oil spill responsibilities and provide direction to the ICT.[73]

At 1600 on Thursday, March 30, Castellina and Liebersbach met with officials from the City of Seward and several Kenai Peninsula area agencies. They discussed forming a MAC Group. Bill Lawrence provided guidance on the relationship between such a group, the ICT, and the Regional Response Team at Valdez. At 1700, Liebersbach briefed the newly formed group to discuss its organization and function.[74]

Castellina chaired the MAC Group. Fitzmaurice took the chair in her absence. Other land managers, and representatives of other interested organizations, joined her on the committee. The initial goal was to have ten agencies represented on the committee.

PROPOSED ORIGINAL MEMBERS
OF
SEWARD MULTI-AGENCY COORDINATING GROUP

Alaska Department of Fish and Game
Alaska State Parks
City of Seward
Chugach Alaska
Cook Inlet Aquaculture Corporation
Exxon
Kenai Peninsula Borough
National Park Service
North Pacific Fishermen's Association
U.S. ·Fish and Wildlife Service.

Figure 2-1[75]

Of the ten agencies, only the Fish and Wildlife Service was reluctant to participate. At the time, Seward appeared to be more than one-hundred miles from the spilled oil. Eight days were to pass before oil began to come ashore in Kenai Fjords

National Park and on the nearby Chiswell Islands, a part of the Alaska National Maritime Wildlife Refuge.[76] The Fish and Wildlife Service seemed to lack a sense of urgency then, and later, about participating in the MAC Group. Dave Patterson did not join the MAC Group as the official Fish and Wildlife Service representative until April 12.[77]

Individual members of the MAC Group contributed in different ways, depending upon their backgrounds and their agencies' interests. Don Gilman, Mayor of the Kenai Peninsula Borough, was particularly significant. The borough, comparable to a county government in other states, had local political jurisdiction over much of the coastline affected by the oil spill. Gilman, a retired teacher and school administrator, lived in Seward at one time. Based in Soldotna, the borough seat 80 miles northwest of Seward, at the time of the spill, he was on his second term as borough mayor. He had also been a state legislator.[78]

Anne Castellina, Superintendent of Kenai Fjords National Park, conducting MAC Group meeting in Seward. (NPS photo.)

Gilman brought decisiveness, personal energy, political
savvy, and a far flung network of political acquaintances to his
seat on the MAC Group. In Gilman's words:

> We didn't sign anything, we didn't sign any agreements,
> we just said that we're going to form a MAC group,
> we're going to get started, we're going to cooperate on
> this thing and Dave's [Liebersbach] going to be in
> charge.[79]

Funding Mechanisms

On Sunday, April 2, Gilman and Castellina flew to Valdez to
see Dennis Kelso, Commissioner of Environmental Conservation for
the State of Alaska. As borough mayor, Gilman signed an agreement
with Kelso. It said that the state would reimburse the borough
for up to $200,000 of costs incurred in dealing with the oil
spill. Shortly thereafter, the borough assembly authorized
expenditures of up to $3 million for oil spill response. Then on
Thursday, April 6, Gilman and Castellina flew to Kodiak. There
they met with Exxon officials who agreed to reimburse the borough
for up to $1 million of costs incurred in dealing with the oil
spill. The borough, in turn, agreed to reimburse the NPS for ICT
work on municipal, state, and private lands.

When Department of the Interior officials objected to the
reimbursement arrangements, Gilman contacted the Washington
office of Sen. Stevens. The objections disappeared by ten
o'clock the next morning.[80] Thus, in addition to proving to be
an effective member in the day to day deliberations of the MAC
Group, Gilman led the effort to structure funding that allowed
the ICT to operate in a comprehensive fashion.

Liebersbach came to the first MAC Group meeting on April 2
to explain his team's role, and to explain that its work did not
imply federalization of the incident. Only federalization would
bring federal dollars. In the meantime, MAC Group participation
did not require financial contribution. But available federal
funding could be used only for work on Park Service land, while
the state channeled funds through the Kenai Peninsula Borough for
work on state, local government, and private lands.

Jack Sinclair, a seasonal ranger at Caines Head State
Recreation Area, represented Alaska State Parks on the MAC Group.
Caines Head is ten miles south of Seward on the west shore of
Resurrection Bay. Sinclair volunteered to do a critical resource
inventory for all of Resurrection Bay with assistance from the
State Department of Fish and Game.[81] The inventory, later used
to support boom deployment priorities, exemplified cooperation
engendered by the MAC Group.

Daily Activities

The newly formed group determined to meet daily at 0900 at
the Kenai Fjords National Park Visitors Center. The physical

33

characteristics of the meeting room influenced the way the MAC Group conducted its business. A large, long conference table almost filled the narrow meeting room. The space available permitted a single line of chairs against each wall behind MAC Group members clustered on two sides of the table. The meetings were open to the public, but the constricted space in the meeting room limited attendance. This kept each meeting from becoming a contentious "mini public hearing," something that happened to a similar group that later functioned in Homer.

From the outset in Seward, Castellina structured the meetings so that the members could discuss many subjects in a short period of time. Typical agendas included a number of standardized topics. Most meetings began with a review of the previous day's session. Then, necessary because agency represen- tatives and others in attendance changed frequently, Castellina defined the group's membership. She then obtained consensus on boom deployment priorities. Information items followed before the MAC Group considered requests that would be forwarded to the ICT for action. Other needs upon which the group agreed were taken up as work items by various MAC Group members.

Beginning with its initial meeting, the MAC Group demon- strated concern that the public know about oil spill response activities. On April 2, the group directed that the ICT public information officer issue daily briefings. The following day, the MAC group supplemented this. It ordered that copies of the briefings be sent by facsimile transmission to all towns on the Kenai Peninsula.[82] Later, after Castellina and Gilman had observed a public information meeting in Kodiak, the MAC Group asked the ICT to set up such a meeting in Seward on the evening of Saturday, April 8.[83] These tasks and a constant stream of telephone and in-person inquiries about spill response kept ICT Information Officer Dixie Dies and rotating public affairs officers from the Bureau of Land Management, Forest Service, and NPS constantly busy.

In addition to issues that directly concerned the ICT and NPS, the MAC Group addressed related subjects. These included the need for decontamination stations to treat vessels that had moved through the oil slick, wildlife rescue centers, methods for breaking up the oil slick, skimming activity, storage of oil killed carcasses, and waste disposal. Speaking as the voice of several entities, the MAC Group had a combined strength that exceeded those of lone voices.

Exxon, with a seat at the MAC Group table, reimbursed local governments for expenses approved by the MAC Group.[84] The oil company also heard from the MAC Group when members believed it was too slow in responding to oil spill needs. Delays in setting up bird and sea mammal rescue centers, and in sending personnel to Seward figured prominently in the group's complaints to Exxon.[85]

34

Homer MAC Group Advisory Committee

With oil spill response activity centered in Prince William Sound and in Seward, coastal communities west of Seward clamored for attention. Homer residents feared that the oil would affect their fishing grounds and the salmon streams that fed them. When this resulted in an ICT branch in Homer, Gilman, Liebersbach, and Douglas D. Erskine determined that activity in Homer would be overseen by the MAC Group in Seward.[86] Erskine served as an Alaska Regional Office, NPS, liaison with the ICT. He was Fire Management Officer for the Alaska Region. Detailed to the Interagency Fire Center in Boise as the Park Service's Acting Chief of Fire Management when the Exxon Valdez spill occurred, Erskine had been involved at that level in mobilizing the ICT. A park ranger for 28 years, Erskine had extensive experience with the Incident Command System.[87]

As a result of the conversation between Gilman, Liebersbach, and Erskine, Castellina proposed that the Seward MAC Group accept a plan to set up a "Mini-MAC" in Homer to set priorities for that area.[88] Later this idea was refined, with one person from Homer, advised by a committee of 12 in Homer, being seated on the Seward MAC Group.[89] The Homer committee's area of responsibility extended from Anchor Point, 14 miles northwest of Homer on Cook Inlet, south and southeast to Gore Point at the western limit of Kenai Fjords National Park on the outer Kenai Peninsula coast.[90]

The plan to have the Homer MAC Advisory Committee work through the Seward MAC Group soon created difficulties. Communications between the two bodies first had to be extracted from hastily written minutes, then sent by facsimile. Often the communications themselves did not accurately reflect the intent of the originating body. Even when they did, the receiving body sometimes misunderstood what it had received. Efforts by the ICT to run an aircraft shuttle between Seward, Soldotna, and Homer for key personnel such as Gilman and Loren Flagg, chair of the Homer committee, helped only a little. The time required for travel ate into the travelers' already overfilled days. When participants in the Homer area failed to develop the cohesive approach to oil spill response adopted in Seward, this compounded communications problems between the Seward MAC Group and the Homer MAC Advisory Committee.[91] Ultimately the Homer MAC Advisory Committee functioned independently. The fiction of its ties to the Seward MAC was maintained because the Incident Command System did not allow for two MAC Groups functioning simultaneously.

35

When two additional NPS units, Katmai and Lake Clark, requested ICT assistance the MAC Group endorsed the requests. It also recommended that response activity extend to state coastline sandwiched between the two national areas. In doing so, the MAC Group acknowledged that this work could not be federally funded. At the same meeting, on April 8, the MAC Group informed DOI representatives at Valdez that its (the Seward MAC Group) area of concern ran "from Resurrection Bay to Katmai National Park and Preserve inclusive of state, local, city, federal, borough and private lands with the exception of Kodiak."[93]

Transition to ICT Phase-Out

When the need for the ICT seemed to be diminishing and the standard twenty-one day rotation of the team was expiring, the MAC Group developed requirements that Exxon needed to meet in order to take over response management from the ICT. The group approved Exxon's participation in unified command and asked that Exxon take over responsibility for deploying and maintaining boom. Exxon did so on April 14.[94] The coordinating group also tasked Exxon to provide a daily operational plan similar to that produced by the ICT.[95] Finally, the MAC Group directed that cleanup activities would not take place without its approval.[96]

Coast Guard Coordination

Late on the evening of March 30, at 2100, Capt. Rene Rousell, U.S. Coast Guard, met with Castellina and Liebersbach. Rousell was both Commanding Officer of the Coast Guard's Marine Safety office in Anchorage and Assistant Federal On-Scene Coordinator (FOSC) for Western Alaska. As such, he held authority for direction of federal oil spill response outside Prince William Sound. He stressed that it was his responsibility to coordinate the federal spill response in Seward. In discussing potential impact of the oil on Kenai Fjords National Park, Rousell emphasized the need to identify and establish priorities for areas to be protected.[97]

Capt. Rousell came to the Exxon Valdez spill with extensive related experience. During his seven years as a commanding officer, he dealt with many spills in Florida and in Alaska. One occurred in July 1987, when the tanker Glacier Bay spilled oil into Cook Inlet to the north and west of Seward. Although of less magnitude, this spill was very similar to the Exxon Valdez incident. Rousell knew that North Slope crude oil, spilled both from the Glacier Bay and the Exxon Valdez, would very quickly turn into a thick, pudding-like substance called "mousse." The mousse, a deadly mix of weathered, thickened oil, debris, vegetation, and the carcasses of the oil's victims, would be very difficult to protect against or capture. The equipment even for attempts to do so was not, Rousell knew, available in Alaska.[98]

Peppery and energetic, Capt. Rousell faced two major problems and thought he might have a third. The first was the task of getting Exxon, the oil spiller, to extend its spill reaction activities beyond Prince William Sound. The second was to mobilize Coast Guard resources to deal with the spill outside Prince William Sound.

The first problem proved difficult to resolve completely. Rousell met with initial success in dealing with it, but had to prod Exxon continually during the rest of the spill response work.[99] The second problem was more easily solved. A few days after the spill Rousell was able to obtain additional personnel to establish a Coast Guard presence in Seward, Homer, and Kodiak. Coast Guard personnel also established portable weather stations outside Prince William Sound. Spill reconnaissance flights, called Air Eye, increased. Coast Guard vessels took on a variety of tasks.[100]

The third problem turned out to be one of perceptions: by Rousell, quickly abandoned, that the NPS wanted to assume the role of FOSC in Seward; by the Park Service, also quickly abandoned, that Rousell wanted to squash its response activities. These soon abandoned perceived problems resulted from an unprecedented situation. In most situations, the Coast Guard would have initiated the activities started by the Park Service. Circumstance gave that role to the Park Service in Seward. Rousell at first considered the ICT a "loose cannon" in an oil spill response situation.[101] With some time to assess what was going on, the Coast Guard joined in ICT operations and MAC Group deliberations while retaining its role as final authority over response work.[102]

By the morning of April 4, the Coast Guard had changed its attitude toward the ICT. On reflection (in speaking of the Seward ICT), Capt. Rousell told the federal Regional Response Team in Anchorage: "It's perfect."[103] He covered his change of mind gracefully. Capt. Rousell attributed the change to Sen. Stevens. According to Rousell, the senator told the Commandant of the Coast Guard he was not impressed with the Coast Guard operation in Valdez. The "Coast Guard's" Incident Command Team in Seward had impressed him. Stevens advised the commandant to keep up the good work in Seward.[104] Later, at a core team meeting, Liebersbach told his key staff that Stevens had known that the ICT was not a Coast Guard operation. He thought Steven's conversation with the commandant had been a tactful way of supporting the ICT.[105]

Expanded Scope of Operations

Boom Deployment

National interest extended to the Seward operation. It encouraged the ICT to extend the scope and nature of its activities to include work outside National Park areas and to include boom deployment. Senator Stevens visited over the April 1-2 weekend.[106] He provided significant encouragement and advice. According to Castellina,

> Stevens said: "you know, you guys are going to get hit." And he was the only one in those early days who believed that. The only one outside our own Park Service people and the City of Seward who believed that.

The senator also advised Castellina and Liebersbach not to be deterred by the fact that Exxon or the Coast Guard were not in Seward. He encouraged them to do all that needed to be done to protect the resources of the area.[107]

Stevens brought with him a map prepared by Dr. Tom Royer, a University of Alaska oceanographer. Royer's map predicted that the oil spill would soon strike the coast of the Kenai Peninsula.[108] Castellina and Liebersbach were debating whether or not to start defensive booming. They were inclined to do so. Stevens' comments left them with no hesitation.[109]

On April 3, the ICT integrated the City of Seward into a unified command structure. Seward Emergency Operations Chief John Gage joined Liebersbach, serving as co-commander of the team.[110] After Stevens' visit and initiation of unified command, the ICT's responsibilities became four-fold: (1) collecting current intelligence to help the MAC Group make decisions; (2) dispatching and supporting teams gathering intelligence to support future litigation and management; (3) dispatching and supporting teams to place and monitor defensive booms; and (4) serving as a focal point of spill activity and information for the community. Predicted movement of the oil soon required that these activities also be carried out from other coastal communities west of Seward.

Branch Operations

As the oil moved out of Prince William Sound and down the Kenai Peninsula coast, Katmai Superintendent G. Ray Bane and Lake Clark Superintendent Andrew E. Hutchison knew that their parks were also in danger. Preliminary planning began almost immediately. Daniel M. Hamson and Cordell Roy, Environmental Specialists from the Alaska Regional Office of the NPS, flew to Katmai headquarters at King Salmon to assist Janis M. Meldrum, the park's Resource Management Specialist, in developing priorities for resource protection. Meldrum came to Katmai with experience as a Resource Management Specialist at two other national parks and training and experience with the Incident Command System.[111]

Between April 7 and 12, Hamson, Meldrum, and Roy wrote the Katmai oil spill contingency plan. They also, by telephone, organized pre-oiling assessment teams to survey the coastlines of Katmai and Aniakchak.[112] During this time, Superintendent Bane flew to Seward, Homer, and Kodiak to assess potential bases for park protection efforts. He, along with Hamson and Roy, determined that for a number of reasons Kodiak was the most logical base of operations. At Kodiak NPS staff would have access to a wide range of vessels. There was good air access directly across Shelikof Strait to any point along the Katmai coast, as well as opportunity to establish good communications to the coastline. Homer was just too far away.[113]

Hamson, Meldrum, and Roy recommended that the Katmai response be a mirror image of the Kenai Fjords response. They identified key resource areas that should have high priority for pre-oiling assessment and for protection. The trio's personal knowledge and information gleaned from prior bird and mammal surveys provided the basis for their recommendations. The recommendations, if accepted, meant that the enormous drain on Park Service resources both in the Alaska Region and nationwide caused by the Kenai Fjords pre-oiling assessment and protection activity would continue and expand.

Bane, who had begun working for the NPS as an anthropologist in the Arctic and had been Management Assistant at Northwest Alaska Areas before becoming Katmai/Aniakchak superintendent, was a fierce advocate for his parks. He adopted the recommendations of Hamson, Meldrum, and Roy, then flew to Anchorage to negotiate the necessary resources with regional officials. A Katmai pre-oiling assessment operations similar to that at Kenai Fjords began.[114]

KATMAI RESOURCE RISK ASSESSMENT

Priority	Location	Estimated Boom Needed	Resource Values
1	Geographic Harbor	1700 feet	salmon spawning bear habitat seabird concentrations bald eagle nesting commercial fishing
2	Big River	500 feet	
3	Swikshak Lagoon	1000 feet	
4	Ninagiak Lagoon and River	500 feet	
5	Takli Island	2000 feet	
6	Kaflia Bay	1200 feet	
7	Chiniak Lagoon	800 feet	
8	Dakavak Lagoon and River	200 feet	
9	Kukak River and Bay	2000 feet	

Figure 2-3[115]

On April 4 the Katmai staff released a list of areas to protect. Naming the areas in priority order, the list also estimated the amount of boom needed to protect them.

Remote from the nearest community, Katmai did not enjoy the benefits of proximity which were available to Kenai Fjords from the City of Seward and the Kenai Peninsula Borough. Kodiak, however, proved to be a good base of support.

In Kodiak the Emergency Service Council, an organization similar to the MAC group in Seward, was already in place. The National Park Service joined the group in early April and was instrumental in the establishment of the Kodiak Inter-Agency Shoreline Cleanup Committee (KISCC). KISCC played a significant role in establishing priorities. Prior to the establishment of a NPS presence in Kodiak, the group maintained contact with Bane and Meldrum in King Salmon through telephone conferencing. KISCC represented Katmai and Aniakchak and fought for park priorities.[116]

Before the oil spill, park management had scheduled visits to Kodiak to meet with residents and discuss park management, but the visits never materialized. Positive long term contacts with

Katmai National Park
and Preserve

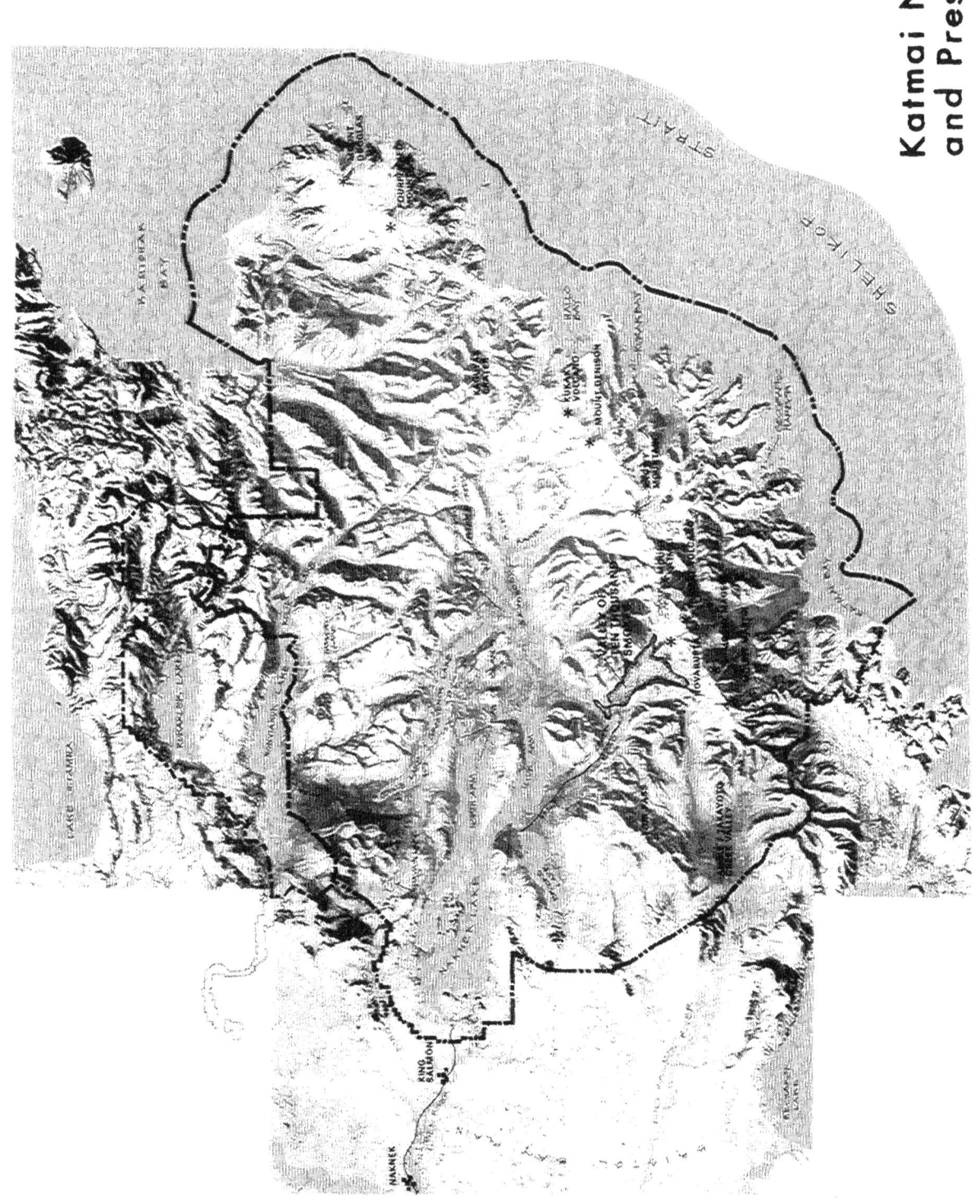

953 | 80,118
FEB 90 | ARO

the Kodiak borough and local fishermen resulted from park participation in the KISCC.[117] According to Bane:

> When I arrived there in early April, those people
> adopted me. They took me in, they gave me a place to
> work, they offered whatever assistance they could....
> the people of Kodiak welcomed us with open arms.[118]

Still, Katmai/Aniakchak did not have the advantage of the long-term, established relationships in the community that benefited Kenai Fjords so greatly. There was no non-federal entity to buy boom. Bane finally convinced the Coast Guard and emergency response officials in Kodiak to allocate some boom to Katmai. But by the time the allocation was made, and necessary coordination accomplished, oil had begun to flow through Shelikof Strait and wash up on Katmai's beaches. Although many parts of the Katmai coastline were exposed and not appropriate for booming, some areas such as tidal lagoons and the inner parts of indentations such as Swikshak Bay might have been partially protected. This could have kept some oil out of critical nesting habitat and bear habitat.[119]

The day after Katmai submitted its protection priority list, Hutchison advised Liebersbach that his park "would like to participate in the current incident management program to protect resources within Lake Clark National Park." Lake Clark's first request was for a photography team to document the coastal ecosystem.[120] The MAC Group then directed the ICT to develop a plan for branching the team in case action was necessary outside the original control area.[121]

At 1000 on April 6, Liebersbach met with his staff and NPS officials to establish an ICT Branch at Kenai, a small town on the northwest side of the Kenai Peninsula. The Kenai operation would coordinate pre-impact intelligence gathering in Lake Clark National Park and Preserve. Liebersbach named Tom Goheen as director of the newly established branch. Jim Ward, originally Air Operations Director for the ICT, substituted for Goheen as Operations Chief at Seward. Discussions later that day between Liebersbach, the MAC Group, and NPS covered how work planned for Lake Clark National Park and Preserve, Katmai National Park and Preserve, and Aniakchak National Monument related to the MAC process. While the discussions were going on, the ICT received a message from Chief Ranger Stephen M. Hurd, Acting Superintendent at Katmai in Bane's absence. As a result of telephone contact with Bane, Hurd requested ICT assistance for Katmai and Aniakchak.[122]

Bane, like Castellina, had concerns about calling in the ICT. He anticipated complications from bringing in an emergency response structure different from the one already operating in Kodiak.[123]

That evening, Liebersbach, Castellina, Rousell, and Gilman met to discuss ICT/MAC involvement in Homer and Kodiak. Gilman, concerned to assist his constituents in Homer, urged that an ICT branch be established there. The MAC Group, which had discussed the issue at its 0900 meeting on April 6, supported this recommendation. Vessels available for charter at Homer were more suitable than those at Kenai for the rough water operations that would be necessary to send field parties to the Katmai coastline. Air operations to Katmai would also be better staged from Homer or Kodiak.[124]

Following the evening meeting, Liebersbach gathered his staff to plan for ICT branch operations in Homer.[125] At the meeting, the core staff discussed the work that needed to be done at Homer, how to do it, and the tense political situation there. Homer residents, frantic over the danger to their fishing industry livelihood, were demanding but not getting action from Exxon. In the end, the core team decided that the ICT would maintain its Seward operation and have two branches: one in Kenai and one in Homer.

The Seward ICT would continue to service Kenai Fjords National Park and also coordinate activity of the two branches. The Kenai branch would work in Lake Clark National Park. The Homer branch would work in Katmai National Park and Preserve and Aniakchak National Monument, although Superintendent Bane continued to insist it would be better done from Kodiak.[126] It would also work on state, local, and private lands in the Homer area. Joseph P. Stam, one of Liebersbach's operations deputies, would go to Homer as branch director there. Overall, it appeared that the Homer branch would serve a political, rather than an operational need.[127]

The ICT also learned that Wrangell-St. Elias National Park, to the east of Prince William Sound, anticipated a limited need for ICT assistance in about a week. This never materialized, although the Regional Office did send staff to the park to complete its oil spill contingency plan and assemble resource information.[128]

The ICT branch at Kenai opened for business on April 7 in the Lake Clark National Park and Preserve offices in Kenai. The ICT branch at Homer was operational that evening, having established a command post in unused bar and restaurant spaces at a motel in Homer.[129]

About the time the ICT set up in Homer, arrangements were made to put a Katmai Superintendent's Representative in Kodiak to coordinate park related work of the emergency response center there. The small Katmai/Aniakchak staff (eight permanent employees at the beginning of the spill) could not be spared for long assignments away from park headquarters at King Salmon. Regional office staff and later NPS staff from other regions represented Bane at Kodiak, although spill concerns and demands continued to dominate his schedule.[130]

On April 8, the Homer branch ICT held an open meeting to explain its purpose to the general public. There was a lot of confusion about that purpose, both within the branch staff and on the part of the public. In the end, the branch defined its function as "to support agencies with jurisdiction over [the] incident." Daily public meetings, held at 1100 in the Homer City Council chambers, followed this first meeting, as did a meeting in Seldovia, a fishing village across Kachemak Bay from Homer. Later, on April 11, there was a public meeting in another village across the bay, Port Graham. That evening Homer was the scene of still another public meeting. It was, according to Homer branch ICT officials, poorly organized. There was little ICT involvement at the meeting. Many of the 220 participants accused Exxon of failing to meet Homer's needs.[131]

Fearful of what the oil might do and frustrated by a seeming lack of response, the Seldovians requested an Incident Commander, Public Information Officer, and Finance Chief in their community. Referred to the MAC Group in Seward by Liebersbach, the request was not fulfilled.[132]

While the Seldovians did not get the ICT they requested, the U.S. Fish and Wildlife Service at Kodiak got an ICT it didn't want. In response to a request from Department of the Interior officials in Washington, Liebersbach arranged for a Type-II ICT headed by Dave Dash to go to Kodiak. The Type-II Team functioned independently of the ICT at Seward. When the Kodiak team was in place on April 10, the Homer Branch ICT relinquished responsibility for Katmai and Aniakchak operations from the Homer Branch ICT to Kodiak.[133] Goheen initially went to Kodiak "to work with but not for" the new ICT in coordinating Katmai and Aniakchak activities. This soon changed. Goheen demobilized. The Kodiak team assumed full responsibility for oil spill response in the NPS units north of Kodiak Island.

A decision by Exxon and the Federal On-Scene Coordinator to establish the command center for the Kodiak Sector (approximately one-third of the total spill area) in Kodiak was responsible for the decision to establish a separate NPS field office there. Facilities were acquired, communications and air and sea transportation systems established, and management infrastructure developed to support a staff of forty for the summer-long response and damage assessment activities.

The transfer of responsibility for Katmai and Aniakchak operations, plus the increasing presence of Exxon personnel in Homer, started planning for closing the Homer branch ICT. April 14 became the target date for demobilization.[134] About the same time, officials at Lake Clark National Park and Preserve decided their objectives had been met. The Kenai branch ICT scheduled itself to close down on April 13.[135]

The Kenai Branch ICT demobilized its personnel and delivered its equipment to Homer. The Homer Branch ICT demobilized more gradually. Exxon personnel, who first arrived in Homer on April 10, initially staffed the command post's reception area, then participated in briefings, and started to run the daily public meetings. By April 14, Exxon had almost totally taken over the ICT's functions in Homer. Although Stam, the Branch Director, remained at Homer until April 17 to assist Exxon, the branch itself closed down on April 16. Garey Coatney, Chief of Land Resources Division for the Alaska Regional Office, who was

44

detailed as Plans Chief for the Homer branch, remained in Homer until April 29. He and Brad Cella, Resource Management Specialist for the Alaska Regional Office, debriefed the last of the intelligence gathering teams as they returned from the Katmai and Aniakchak coastlines.[136]

Comment

The Incident Command Team in Seward had worked well. As later events would prove, the Kenai Fjords National Park staff was too small to have successfully managed the influx of people and avalanche of requirements that were necessary to collect data, deploy booms, and coordinate plans before the oil hit the park's coastline. The MAC Group, too, had worked well. Supported by the ICT staff and drawing upon Castellina's skill as chair and the good relations she had established with the community prior to the incident, the group proved to be an effective means of setting priorities and coordinating activity of a mix of federal, state, local and private entities.

The two branch operations of the Seward ICT had varying success. The Kenai branch, serving the limited needs of Lake Clark National Park, was able to complete its tasks efficiently. Unlike the Homer branch, Kenai had the additional advantage of not working in an environment of civic turmoil.

The Homer branch ICT faced a number of problems over which it had no control. By the time it arrived in Homer, local residents were already impatient with what they viewed as government inattention. While the branch ICT might have rectified this, it suffered too from a lack of cooperation by field offices of federal and state agencies. Homer offices of the state Department of Fish and Game, despite Schroeder's earlier participation with the MAC Group in Seward, acted independently. At one point, Fish and Game employees hijacked boom intended for other destinations and took it to their fish hatchery outside Homer.[137] The MAC Group Advisory Committee in Homer, even after being freed from its ties to Seward, never established the control and public acceptance that were achieved by the Seward MAC Group. As a result, much of the Homer committee's meeting time was spent addressing concerns raised by angry members of the public attending its meetings.

Although the Homer branch ICT was able to dispatch intelligence gathering teams to the coastlines of Katmai and Aniakchak, it never achieved the leadership role attained by the Seward ICT. The political situation in Homer surpassed the political situation in Seward in complexity. Had the resources been available to do so, a separate Type-I ICT might have better dealt with the situation in Homer. Despite the desirability (in retrospect) of doing this, a second Type-I Incident Command Team was not called. Of the 18 Type-I teams, the Service had mobilized one for a non-fire incident at Fire Island National Seashore in New York, had a second Type-I team in Seward, and hesitated to call in a third Type-I team for a non-fire incident.

45

Even the use of two Type-I teams for non-fire incidents caused some grumbling by other agencies participating in the Incident Command System.

Criticism of Park Service use of the two Type-I ICTs was aggravated when the Kenai Peninsula Borough called in a third Type-I team to operate in Homer after the Homer branch operation of the Seward ICT shutdown.[138] The difficulties that this attitude created pointed out the need for all agencies participating the Incident Command System to accept the use of ICTs in non-fire emergencies.

The final test of these arrangements for command, control, and coordination lay in their result. These were the staff and field operations undertaken to prepare for the oncoming oil.

CHAPTER 3 - STAFF AND FIELD OPERATIONS

- Overview
- ICT Staff Operations
- Field Operations
- Comment

Overview

Staff and field operations tested the command, control, and coordination mechanisms developed to manage NPS response to the _Exxon Valdez_ oil spill. Staff operations included functions such as Planning, Logistics, and Finance that supported the field operations of pre-oiling intelligence gathering and booming.

ICT Staff Operations

Set-Up

After setting up in Seward in the small house owned by the U.S. Forest Service, the ICT prepared its first "shift plan," to guide action on April 1. That first plan established objectives of surveying marine animal populations and bird staging areas on land and water, establishing priorities for protecting areas from Bear Glacier south on the coastline; identifying areas where booms could be used; and developing a unified interagency organization. While the ICT developed these new intelligence gathering teams to survey coastline, two other teams were in place.

One of the in-place teams was a photography group. John C. Black, Manager of the Department of the Interior's Training Center at the Boise Interagency Fire Center, and a professional photographer, led the group. In the field on March 31, the team traveled by 42-foot boat to collect video and still photographic documentation of the Kenai Fjords coastline.[139] Photo team members later photographed headquarters staff, intelligence gathering, and boom deployment teams as they went about their business.

The Park Service's Civil Litigation (or Tort) Team was the other in-place group. Led by Leland J. "Lee" Shackleton, the Tort Team established a chain of custody for collected data in anticipation of its use as evidence in litigation. Team members, all trained in law enforcement, also debriefed other ICT personnel as they returned from field work.

For the other four teams, Bud Rice developed a data collection plan that required investigators to debrief with the ICT's Plans Section before being demobilized. The debriefing included turnover of complete sets of data, including maps.[140]

Daily Operations

As a part of the second day's operations, the ICT identified classes of specialists it would need to carry out its mandate. Orders placed through the Incident Command System started to produce results. Experts in various fields began to report in to Seward.[141]

Subsequent daily operations followed much the same pattern as those of the first day. Mapping the oil slick and establishing reliable communications with field crews became additions to the daily routine.[142] Standard events characterized each day. The first daily event was an 0800 briefing for all team members. At these briefings, Anne Castellina, the Incident Commanders, and Sections Chiefs reported on the current situation and outlined planned activities for the day. The core team members, Dave Liebersbach, Don Fuller, Tom Goheen, Ron Knowles, and Marv Robertson were conspicuous at the briefings and throughout the day because of their yellow shirts, a fire-fighters' uniform.

At the end of each day, an 1800 general briefing updated everyone involved on the progress of that day's activities, the location of the oil, and impending weather. The general briefings were followed by individual section meetings at which section members discussed work to be done. There were also core team meetings. At these Liebersbach and his key staff reviewed the political situation, which changed from hour to hour, and the overall progress of ICT operations.

Aware of the complexity of the incident with which they were dealing, ICT leaders continually stressed the importance of documentation of "all decisions, action, and considerations."[143] Involvement of several agencies and multiple funding sources complicated accountability. Everyone involved was aware that at some point the federal and other governments would litigate with Exxon Corporation for cost recovery.[144]

By April 3, the NPS had 45 people from the Service and other federal and State of Alaska agencies assigned to the ICT at Seward. Another 25 people in the Alaska Regional Office worked part-time on the "Seward Incident." Regional officials alerted other Park Service regions to a potential call for assistance from other regions. Eventually over 500 NPS personnel participated in spill response, at great cost to park programs and themselves. [See Appendix B.] The greatest impact was on the Alaska Regional Office and the parks it managed, where many activities scheduled for the summer of 1989 had to be postponed or cancelled. Disruption of schedules, long hours, and frequent travel affected all those involved to some extent. Deeply dedicated to the resources with which they worked, personnel in Alaska also found oil's destruction of wildlife and natural beauty to be particularly horrible. Recognizing this, the regional directorate arranged for stress counselling for personnel who wished to take advantage of it[145]

Information from Prince William Sound highlighted the extent of the disaster looming for Kenai Fjords National Park and

Preserve. On-the-ground counts on the sound's shoreline revealed as many as 130 dead birds per mile in one stretch of Montague Island coast. Biologists estimated that dead bird counts might run as high as 250 birds per mile in other areas of the sound.

Planning

Planning for intelligence gathering was initially driven by Rice, as the resource coordinator who identified what needed to be done to document and protect park resources. When ICT activities expanded beyond the boundaries of Kenai Fjords National Park, William B. "Brad" Cella from the Alaska Regional Office joined the ICT to coordinate the input of park resource specialists such as Rice, Janis Meldrum and Dave Manski for Katmai and Aniakchak, and Bob Gerhard for Lake Clark. The MAC Group, collecting and approving booming priorities from a variety of sources, drove planning for boom deployment.

Early attempts to use satellite imagery to locate and track the advancing oil failed. After LANDSAT, an American satellite, wouldn't turn on properly and a French satellite missed its shots, oil spill authorities abandoned efforts to coordinate satellite coverage. Locating the oil depended on aerial observation and imagery.[146]

In addition to Rice, specialists such as Blair Young, the ICT's Situation Unit Leader, and Joe Santa Maria, a boom manufacturer's representative for JPS Incorporated, made periodic flights to check the progress of the oil slick and of boom placement operations. As a result of these flights and with information provided by Exxon, the National Oceanic and Atmospheric Administration, and the U.S. Coast Guard, Young produced daily maps showing the location of the oil as it moved to the southwest and the status of the boom deployment effort. As Resource Unit Leader in the Planning Section, Joe Ribar controlled the assignment of incoming personnel and demobilization of people whose ICT work had been completed. Robertson, as Planning Chief, coordinated the work of the section's units and also individual specialists such as archaeologists, historians, and meteorologists assigned to his section.

Paul Gleeson, Compliance Archeologist for the Alaska Region, NPS, was the first cultural resource specialist to report to the ICT. He had become involved with the spill several days earlier when he worked with Paul Gates and John Mattson, archeologist for Chugach National Forest, to develop the Forest Service's cultural resource response to the spill. The Forest includes much coastline and many islands in Prince William Sound.[147]

Gleeson's first job with the ICT was to establish procedures for identifying cultural resources that might be affected by the oncoming oil. He also alerted the federal spill response manager, the Coast Guard, to that agency's responsibilities under Section 106 of the National Historic Preservation Act. The Act requires that a federal agency take into account the impact of its actions on cultural resources. Gleeson had participated in a

49

joint US/USSR oil spill response exercise. He knew from that experience that major impacts to cultural resources were likely to come not from the oil itself but from cleanup activities. While preparing for this, he arranged for archaeologists to accompany the natural resource pre-oiling assessment teams that worked in Lake Clark, Katmai, and Aniakchak.[148]

Training

Marv Robertson also lectured at sessions held when the Park Service decided to use the Oil Spill Incident as a training opportunity. This was the first time that an ICT managed an oil spill response. Previous uses had been limited to fire attack, law enforcement, and search and rescue efforts.[149]

Art Latterell, a BLM employee, came in to conduct four two-day training sessions. Latterell used a modified version of the Incident Command System (ICS) Course ICS-200 developed by the U.S. Forest Service at the Northern Training Center in Missoula, Montana. NPS, City of Seward, Kenai Peninsula Borough, and Exxon students took classes on April 3-4, 5-6, 7-8, and 9-10. In all, 26 people, including several Exxon employees, received ICS training.

The ICS training began with classroom presentations at the Kenai Fjords National Park Visitor Center. These included 45 to 60-minute talks by Planning Chief Robertson. The first day of training ended with students attending the 1800 briefing at the Incident Command Post (ICP). The second day began with an operations briefing followed by sessions with the various ICT sections. Overflights of the oil spill were included in the afternoons. As a part of their training, students completed ICT check-in and demobilization procedures as part of their in and out processing.[150]

Seizing Seward ICT operations as a training opportunity proved important as spill response staffing requirements expanded and continued. Alaska's regional office and every park in Alaska, plus many outside Alaska, had staff involved in the oil spill incident. This early training allowed many, especially those involved in the early days of the incident, to contribute more effectively.

Logistics

Fuller's Logistics Section did all the things necessary to support ICT headquarters staff and crews in the field. These ranged from housing and feeding support staff to placement, maintenance, and repair of remote communications and weather facilities.

Personnel flowing into Seward soon swamped the small town's hotel accommodations. The Logistics Section booked almost all of the available hotel beds and assigned team members to them as they reported for duty. Fuller also arranged to have local restaurants accept meal tickets as a convenience for team members who arrived in Seward short of cash.

Team requirements for supplies and equipment soon ran into thousands of dollars in costs. Logistics ordered many of the requirements through ICT dispatchers in Anchorage. Other requirements were obtained locally. Fuller's staff distributed them all from a garage adjacent to the Forest Service cottage that had originally housed the ICT. When oil spill response activity overwhelmed the cottage's small rooms, Logistics rented an empty retail store on Seward's main street.[151] It soon became the Incident Command Post not only for the ICT but also for the Coast Guard and the state's Division of Emergency Services.

The ICT supplemented the Coast Guard remote weather stations with its own, placing two Remote Automated Weather Stations (RAWS) in Prince William Sound on April 1. Teams set up the RAWS on Danger Island and on Evans Island, both at the southwest end of the sound.[152] These were the first two outposts in a system of weather and communications facilities that the ICT placed along the track of the oil spill. On April 7, the team set up two radio repeaters on Rugged Island, 19 miles south of Seward in Resurrection Bay, and on the Harris Peninsula, a point of land about 30 miles southwest of Seward that separates Aialik and Harris Bays. The team needed the repeaters to be able to communicate with field parties traveling by boat. Communications with aircraft flying in the farthest regions of the park also required repeater service. Two days later this need led to a third repeater going into operation on Red Mountain, 10 miles southeast of Seldovia.[153]

Additional RAWS followed: on Barwell Island, at the entrance to Resurrection Bay; on Outer Island and East Chugach Island, at the southwest end of Kenai Fjords National Park; and on Marmot Island, a few miles north of Kodiak. Technicians also reinstalled the Rugged Island repeater, which had broken down.[154] By April 11, VHF repeaters were on Ragged Island on the east side of Nuka Bay, on Granite Island 35 miles southwest of Seward at the end of Harris Peninsula, and on Rugged Island. A fourth repeater at Three-Hole Bay, on the west side of the Aialik Peninsula 26 miles southwest of Seward was in place but not operational. Rugged Island was also the location of a repeater serving aircraft.

All the repeaters belonged to the Incident Command System. ICT Logistics Chief Fuller warned the MAC Group that it needed to look ahead to when the ICT and the repeaters would leave the oil spill response.[155]

Finance

Knowles, the ICT's Finance Officer and his deputy, Eva Brown, arrived in Seward on March 31. They immediately set up a Procurement Unit with a Purchasing Agent to handle expenses of up to $10,000 and a Contracting Officer to handle expenses between $10,000 and $25,000. The Finance Section also included a Payment

Team to disburse funds, a Cost Analysis Unit to keep track of expenditures, and a Time Keeper to keep track of the hours worked by ICT members.[156]

A key Finance decision came when Knowles decided to bring in a Payment Team to issue on-the-spot checks to local merchants with whom the Incident Command Team had run up large bills. This reflected the ICT core staff's constant awareness of the importance of community relations. Knowles, and Fuller who also worked closely with local merchants, brought the subject up at almost all general meetings of the ICT. Much of Knowles' personal effort also went to preparing a cost sharing agreement so that the Kenai Peninsula Borough, using state and Exxon funds, could pay for ICT work on non-federal lands.

ICT Field Operations

Intelligence Gathering

Intelligence gathering operations directed by the ICT had three purposes. The most immediate purpose was to provide current information on which the MAC Group could base decisions and that the ICT could use in planning operations. Longer range purposes included compiling data about existing conditions in the park. Park managers needed this information to plan oil spill response and as data to support potential claims under various environmental laws or possible litigation.

Scientists had previously conducted few studies in the park. None provided information on conditions as they might exist in the month of April. New data needed to be gathered to document the park as it existed before the anticipated contamination by oil. The day after it arrived in Seward, the ICT began work to achieve these objectives.

Field Team Structure

The ICT initially structured four teams to collect information. They included two wildlife evaluation teams. One wildlife evaluation team would travel by boat and one would travel by aircraft. A terrestrial evaluation team would travel by boat. A recreation evaluation team would travel by boat.

Gary Vequist, a biologist detailed from the Park Service's Alaska Regional Office, led Team "A," Wildlife Evaluation (boat). The team, consisting of two biologists and a photographer went on the fishing boat Snowbird south along the Kenai Peninsula Coast to Harris Bay. On the way, team members evaluated marine wildlife. Their instructions placed special emphasis on Pederson Lagoon, the moraine area in upper Aialik Bay, Holgate Arm, and the seal pupping area behind the moraine in Northwestern Lagoon.[157]

Mike Nishimoto, a U.S. Fish and Wildlife Service expert on sea birds, led Team "B," Wildlife Evaluation (aircraft). Dale Taylor, another Alaska Regional Office biologist, was the second scientist on the team, which also included a navigator and a photographer. The team included many personnel at various times. Among them were Paul Haertel, Associate Regional Director for Resource Services, Alaska Regional Office, as pilot; Taylor and Nishimoto as observers along with Rice and Chris Titus of Alaska State Parks; and Janet Warburton, a biological technician from the Alaska Regional Office, as recorder. The team, with three observers (including the pilot) and one recorder, made its flights in a Beaver aircraft flying at 500 feet at 80 miles per hour. The team flew first to the Chiswell Islands and Pye Island, then to Nuka Island before flying back up the coast to Resurrection Bay. Enroute and on reaching each specified point, the team evaluated bird colonies and sea lion haul out areas. Between April 1 and 5, Team "B" was able to do four aerial surveys. The surveyed area included all of Resurrection Bay, all fjords and islands of Kenai Fjords National Park and Kachemak Bay State Park. The surveys found the largest concentrations of animals on points or headlands and on islands. There were also a few lagoons that had large concentrations of animals.[158]

Page Spencer led Team "C," Terrestrial Evaluation (boat). The team also included another terrestrial biologist, a fisheries biologist, and one assistant, established test plots to estimate vegetative cover and species identification. The fisheries biologist also took water samples. Priorities for the team's work were Pederson Lagoon, James Lagoon, Delight Creek, Palisade Lagoon, and Beauty Bay.[159]

Bob Gerhard, Management Assistant for Lake Clark National Park and Preserve, led Team "D," Recreation Evaluation (boat). The team consisted of Gerhard, Karen Gustin from Kenai Fjords, Don Dragoo and Belinda Bain, Fish and Wildlife Service biologists. It had two responsibilities: to evaluate areas with high recreation potential and to survey beaches for carcasses of naturally killed wildlife. The team focused on Holgate Arm, Northwestern Lagoon, Pederson Lagoon, Upper Aialik Bay Ranger Station, Delight Bay, and James Bay. Traveling on the M/V Foxy Lady with Captain Mark Bartholemew and deckhand Eric Jackson, Team "D" was in the field from April 2 to 5. It found no naturally killed carcasses and almost no debris on the beaches of Kenai Fjords National Park.[160]

On April 2, a fifth intelligence gathering group, Team "E," Intertidal Survey (boat), joined the structure. Dave Duggins, a biologist on the University of Alaska faculty, led the team, which consisted of himself and one assistant. Team "E's" mission was to survey bivalve animals in intertidal areas and evaluate species composition, distribution, and population density. Traveling on the vessel Kenai Ranger, Team "E" evaluated Pederson Lagoon, James Lagoon, Delight Creek, Palisade Creek, and Beauty Bay.[161]

Incident Commander Liebersbach summed up the intelligence-gathering activity in a press interview:

Our mission is to find out where the marine wildlife and sea birds are today, then set priorities so that the oil that kills them can be contained.... In Yellowstone we put fire fighters on the ground. Here we're providing logistics for Ph.D. biologists who we're handling as a very intelligent resource, very delicately. If they need a boat we get one, if they need a ham sandwich we make one. What they need we try to provide.[162]

By April 4, Teams "A" and "B" were back from the field and working at the Incident Command Post to collate the data they had gathered. Three other intelligence gathering teams that had traveled by boat were back in Seward by April 5. By this time Rice had developed plans for a computer-stored and manipulated data base. Team members responsible for writing data in the field had to transpose their data onto data sheets ready for computer entry before released from the ICT. All had to provide complete sets of data and maps.[163]

While Teams "B," "C," and "D" worked at the ICP organizing information gathered in the field, Vequist took Team "A" back into the field. Duggins' Team "E" continued field work. After resupplying, Team "A" headed southwest to complete its inventory of the still uncontaminated coast.

On April 8, Teams "A" in M/V Snowbird and "E" in M/V Endeavor continued field work, Teams "B" and "C" collated data at the ICP, and Team "D" demobilized. The following day Team "E" finished its work at Nuka Bay and returned to Seward. On April 10, Paul Gabrielson led Team "E" back into the field to work in the Aialik Bay area.[164] By April 11 all teams had returned to Seward and the phase of pre-impact intelligence gathering for Kenai Fjords National Park had ended.[165]

Branch Field Operations

Field operations in Katmai and Lake Clark, directed by the ICT branches at Homer and Kenai, began almost as soon ICT personnel set up the branch offices on April 7. Superintendents at Katmai and Lake Clark advised the Park Service's regional office and the Incident Commander at Seward that their parks needed intelligence collected for a number of areas.

Gerhard left the Seward ICT to become land manager's representative at the Kenai Branch of the ICT. The Kenai Branch formed three intelligence gathering teams. Team "A" led by Richard Harris consisted of two biologists and one photographer. It left Kenai for Homer on the morning of April 9 with the objective of boarding the Bruin Bay, sailing to Chisik Island, and then surveying for population density and species. Team "B" led by Rae Baxter consisted of two more biologists and another photographer. It was to travel by air from Kenai to Tuxedni Bay to survey clam beds near Redoubt Point and tidal flats near the

INITIAL INTELLIGENCE TARGETS
KATMAI NATIONAL PARK AND PRESERVE

Location   Resources

Kamishak River   salmon, bear feeding areas, sportfish
Shaw Island   sea birds, sea mammal haul out areas
Kiukpalik Island   sea birds, sea mammal haul out areas
Shakun Islets   sea birds
Big River   salmon, bald eagle nests, sport fish
Ninagiak Island   sea birds, sea mammal haul out areas
Hallo Bay/Hallo Creek   salmon, bear feeding areas, razor clams
Kukak Bay   bears, bald eagles, sea birds, clams
Amalik Bay   archeological sites, sea birds, sea
                 mammals, bald eagles
Ilktugiak Island   sea lion haul out
Katmai Bay   bear feeding area, razor clams

Figure 3-1[166]

INITIAL INTELLIGENCE TARGETS
LAKE CLARK NATIONAL PARK AND PRESERVE

Location   Resources

Tuxedni Bay   kittiwake rookeries, sea birds,
                        vegetation, clam beds, salmon, migrating
                        waterfowl staging areas, brown and black
                        bear feeding areas
Chinitna Bay   salmon, clam beds, vegetation, migrating
                        waterfowl staging areas, brown and black
                        bear feeding areas, sea birds and marine
                        mammals

Figure 3-2[167]

northern mouth of the bay. Team "C" led by Hollis Twitchel consisted of Twitchel and a note taker. It was to fly from Kenai and survey the park's coastal area for marine mammals.[168]

Superintendent Hutchison accompanied Team "B." Flying from Kenai on the morning of April 9, the team's marine biologists surveyed Tuxedni Bay clam beds while bird biologists photographed bird habitats for number and species. Team "A" went to Homer on the evening of April 9 and set out to cross Cook Inlet. Weather turned back Team "A." Its personnel returned to Homer and then to Kenai on April 10. One Team "A" bird biologist went by helicopter from Kenai to survey Lake Clark beaches while the other team biologist and its photographer went by fixed-wing aircraft on a photography mission. The following day two teams went out for Lake Clark. One, of biologists, took water samples along the coastline. A second team, of another biologist and a cultural resource specialist, flew to Crescent River.

After both teams returned to Kenai and had been debriefed, Gerhard decided that all of his objectives had been met. Snow extending right to water's edge had limited the extent of the Kenai Branch ICT's intelligence gathering. Patricia McClenahan, the cultural resource specialist assigned to the Kenai Branch ICT, went on one of the flights on April 11 and found that a known site, KEN-221 was buried in snow. She made a second flight on April 12 to locate other known sites. This ended activities of the Kenai Branch ICT.[169]

The Homer Branch ICT supported intelligence gathering on the non-federal coastline of the Kenai Peninsula and, initially, along the shores of Katmai National Park and Preserve and Aniakchak National Monument far to the west of Homer. The work on non-federal lands is not detailed here. Concurrent with its planning for demobilization, the Homer Branch ICT began planning to send intelligence gathering teams to the coastlines of Katmai and Aniakchak.

The planners proposed three reconnaissance teams, but feared that enough personnel would not be available. Because of the enormous effort committed to Kenai Fjords National Park at the beginning of April, by mid-April it had been difficult to find qualified personnel to do the brief surveys for Lake Clark National Park and Preserve.

Katmai and Aniakchak surveys, which would require far more time, presented even more staffing problems. Finally, the Homer Branch ICT planned for four teams. Helicopters based in Kodiak and Port Heiden supported the shipborne teams. Teams 1, 2 and 3 surveyed the Katmai coastline and Team 4 surveyed the Aniakchak coastline.[170] Team 4 was unique. It came as a unit from Olympic National Park where its members had gained experience in the December 1988 oil spill off that park's coast.

Homer Branch ICT efforts to request personnel (order resources in the ICS vocabulary) for Katmai and Aniakchak surveys collided with similar efforts initiated by the Type-II ICT at Kodiak. When the ICT dispatchers in Anchorage received duplicate orders they alerted the offices in Homer and Kodiak.[171]

PERSONNEL RESOURCES FOR OIL SPILL RESPONSE
MARCH 31-APRIL 15, 1989

Date & Location	Personnel By Agency									
	AFS	IFC	BLM	DNR	FSV	FWS	NPS	UAK	NWS	TOT*
March										
31 Seward	8	2	0	2	2	0	3	0	0	17
April										
01 Seward	14	2	0	2	2	3	7	1	0	31
02 Seward	14	2	1	2	2	3	9	1	0	34
03 Seward	15	2	1	2	2	3	17	1	0	43
04 Seward	16	2	4	2	3	3	12	1	0	43
05 Seward	16	2	4	2	3	3	18	1	1	50
06 Seward	16	5	4	2	5	1	15	1	1	50
07 Seward	16	5	4	2	5	1	15	1	1	50
Kenai	?	?	?	?	?	?	?	?	?	10
										60
08 Seward	13	3	3	1	5	1	16	0	1	43
Homer	8	0	1	1	0	0	0	0	0	10
Kenai	0	0	0	0	0	0	0	0	0	18
										80
09 Seward	?	?	?	?	?	?	?	?	?	?**
Homer	10	0	7	2	0	0	3	0	0	22
Kenai	?	?	?	?	?	?	?	?	?	18
										73
10 Seward	13	3	3	1	5	1	16	0	1	43
Homer	11	0	6	1	0	0	4	0	0	22
Kenai	4	2	0	0	0	0	5	0	0	11
										76
11 Seward	13	4	3	1	6	1	15	0	1	44
Homer	11	0	5	2	1	0	3	0	0	22
Kenai	4	2	1	0	0	0	6	0	0	13
										79
12 Seward	12	4	2	1	5	1	16	0	1	42
Homer	11	0	5	2	1	0	2	0	0	21
Kenai	3	0	0	0	0	0	5	0	0	8
										71
13 Seward	14	3	2	1	4	1	16	0	1	42
Homer	11	0	6	2	1	0	2	0	0	22
										64
14 Seward	14	3	3	1	4	1	13	0	1	40
Homer	9	0	5	2	1	0	2	0	0	18
										58
15 Seward	12	3	2	1	4	1	10	0	1	34
Homer	2	0	1	1	1	0	0	0	0	5
										39

*(AFS=Alaska Fire Service; IFC=Boise Interagency Fire Center; BLM=Bureau of Land Management; DNR=Alaska Dept. of Natural Resources; FSV=Forest Service; FWS=Fish & Wildlife Service; NPS=National Park Service; UAK=University of Alaska; NWS=National Weather Service); **(? indicates figures not available)

Figure 3-3[172]

Page Spencer taking oil samples in
Kenai Fjords National Park. (Photo
courtesy of Karen Jettmar.)

Chuck Gilbert with dead bird found
on beach at Kenai Fjords National
Park. (Photo courtesy of Karen
Jettmar.)

58

Dispatched on short notice to Homer to embark for the Katmai coast on unfamiliar fishing vessels, the intelligence gathering teams were frantically busy. Once in Homer, the team scientists heard briefings on their objectives, took safety training, and requested special supplies they needed. Logistics staff of the branch struggled to obtain equipment from suppliers as far away as Anchorage, within as little as ten hours turnaround time.[173]

Nancy Deschu, Hydrologist for Alaska Regional Office, led Team 1 of the Katmai pre-oiling assessment surveys. The team departed Homer in the M/V Kittiwake II on April 15 to investigate the Katmai coast from McNeil River to Cape Douglas. It encountered difficulties in operating from a base hundreds of miles distant from its objective. According to Rae Baxter, an intertidal biologist on the team:

> The major problem was the great lack of knowledge about the area or about the job required to be done by most all [Baxter's emphasis] of the people concerned with this study, with the exception of the field crew and the vessel captain who had a little knowledge about the area....Support staff at ICP or where ever [sic] were not knowledgeable about the remoteness of the area, time and tides, and weather conditions. They were apparently unable to interpret the charts and to realize the extent of the intertidal zone and thus the transportation limitations. There were helicopter scheduling problems in that I was never able to get it early enough to be able to work on the low tides.[174]

Despite the difficulties, the team completed much of its work before returning to Homer on April 29. Team 2, led by Dennis Knuckles, sailed from Homer on April 15 in the M/V Stormbird to investigate the Katmai coast from Kukak to Kiupalik. Before Team 2 completed its work, ICT officials recalled it to Homer in the face of severe storm warnings for the area in which the vessel was operating. Team 3, led by Mark Schroeder, sailed from Homer on April 18 in the M/V Widgeon to investigate the Katmai coast from Kinak Bay to Kashvik Bay. The team returned to Homer on April 28. Team 4, led by Douglas Houston, sailed from Homer on April 19 in the M/V Polar Star, to investigate the Aniakchak coastline from Amber Bay to Kujulik Bay and returned to Homer on April 29.

Boom Deployment

Simultaneous with the pre-oiling intelligence gathering in the four park units, the ICT also supported booming. With a minor exception, this occurred only in one park, Kenai Fjords. As part of the cooperative effort through the MAC Group, however, the ICT did support booming on non-park lands.

ICT teams placed booms almost as soon as Castellina and Liebersbach decided to authorize booming. Seward city officials had begun working with Exxon to locate boom on March 29. Exxon purchased the boom and flew it to Seward. It arrived there on the evening of Friday, March 31. On April 1, the ICT Planning Section, using information provided by Bud Rice and Tom Schroeder, determined appropriate boom locations.[175]

On April 2, the first 500 feet of boom went into Humpy Cove. The cove, outside the boundaries of Kenai Fjords National Park, is the outlet of a small salmon stream running into Resurrection Bay. Simultaneously, the ICT prepared to place boom in Thumb Bay, also outside park boundaries.[176]

Joe Santa Maria worked with ICT members on April 3 to install 1100 feet of 36-inch curtain boom in Thumb Bay. On the same day, booming began at Delight Creek in McCarty Fjord. The booming crews didn't finish because of problems experienced with tides. McCarty Fjord is on the outer Kenai Peninsula Coast, within the boundaries of Kenai Fjords National Park.[177]

Determining Priorities

Besides proposing additional members, at its April 3 meeting the MAC Group adopted booming priorities as: (1) Pederson Lagoon; (2) Tonsina Creek; (3) James Creek; (4) Island Creek; (5) Delight Creek; (6) Port Dick; (7) McCarty Lagoon. The group recognized that James Creek, Island Creek, and McCarty Lagoon would require 36-inch or heavier boom.[178]

The group also learned of a $200,000 state fund set aside for oil spill expenses. The Alaska Division of Environmental Conservation established a protocol under which the city needed to obtain MAC Group approval of expenditures before those expenses would be reimbursed from the state fund.

As a part of its April 3 meeting, the MAC Group directed the ICT to locate larger size boom needed for James and Island creeks. The group also ordered the ICT to order absorbent materials needed for potential cleanup.

Coast Guard Authorization of Booming

By the evening of April 3, the ICT learned that the Coast Guard was considering imposing restrictions on boom deployment. Because the Clean Water Act placed the Coast Guard in charge of oil spill operations, a Coast Guard ban would mean the end of booming efforts. All federal agencies had to obtain Coast Guard approval before deploying boom. The Coast Guard thought the MAC Group should give its priorities to the Coast Guard, which would determine where deployments of boom would take place.[179] At the 1800 briefing on April 3, Liebersbach said that the Coast Guard might restrict booming. All federal agencies had to obtain Coast Guard approval prior to boom deployment.[180]

At an evening meeting of the ICT leadership on April 3, John Gage announced that the City of Seward, not subject to Coast Guard restrictions, intended to acquire and place booms. This was particularly urgent. Boats deploying boom had to depart Seward at midnight on April 3 in order to reach proposed boom locations at the right time for tides on April 4.[181] The increased emphasis on booming led the ICT to establish another unit, Team "F," led by Scott Ransom to deploy and monitor booms.[182]

Capt. Rousell had the authority to control boom deployment. Rousell knew that defensive booming was not a widely used concept in catastrophic spill situations of the magnitude created by the grounding of the Exxon Valdez. He decided that while the Coast Guard would not be proactive in booming, boom placement would be good activity for local residents, who could feel they doing something useful. Rousell concentrated Coast Guard efforts on tracking the spill and capturing oil while it was still afloat.[183]

Additional Boom Priorities and Placement

By April 4, Team "F" had deployed boom within Resurrection Bay at Tonsina Creek, Humpy Cove, and Thumb Cove; and outside Resurrection Bay along the Kenai Peninsula Coast to the west at Pederson Lagoon in Aialik Bay. These were the highest priority areas designated by the MAC Group.[184]

When considering additional booming priorities at its April 4 meeting, the MAC Group desired booming at Delight Creek, James Lagoon, Island Creek, Port Dick Creek, Middle Creek, and McCarty Lagoon. Petroff Glacier appeared briefly on the list of priorities, but soon disappeared. The shortened list required 5,700 additional feet of boom. Although 6,000 feet of boom was enroute to Seward, Cal Sikstrom, the Exxon representative to the MAC Group, urged the group to practice "smart" booming. Only essential streams and not bays should be protected. Boom supplies were limited. Exxon, scouring the North American continent and the world for additional boom, was having difficulty obtaining adequate supplies of the barriers.[185]

Peter Fitzmaurice, chairing the MAC Group meeting in Castellina's absence, noted that booms were not totally effective. Some oil could be expected to flow over and some under the booms. He noted that consideration should be given to enhanced protection, perhaps placement of absorbent materials, in sensitive areas.[186]

On April 5, the tender Barlow with 5,700 feet of boom stood ready for deployment to prioritized streams and lagoons. Then the ICT decided to keep the Barlow in port until adequate boom had arrived to protect Resurrection River.[187]

In the end, the MAC Group decided that James Lagoon should be a priority for boom placement. Boom there could protect both fish and sea mammals. It was one of the few areas where boom could protect sea mammals. Other locations, such as haul-out areas, were usually too exposed to heavy wave action for booming to be effective. When the MAC Group discussed this, Dr. Ron Goodman, an Exxon consultant, cautioned against high expectations of success with booming. He noted that the best success came with use of multiple booms, but that the supply of available boom was limited.[188]

The discussion of booming ended with a plea from Dave Firth, a resident of Day Harbor (to the east of Resurrection Bay) for boom protection there. Committee members explained that they had given priority to protection of fish spawning streams and wildlife. It was unlikely that anything would be done to protect Firth's wilderness home.[189]

Disappointment came on April 6. Workers at Boston's Logan Airport had loaded soiled, used, mixed sized boom onto an Alaska Air National Guard C-130 flown across the country to pick up supplies. When the additional boom arrived in Seward at 0100 on April 6, it was unusable. Team "F" learned that it would have to wait 36 hours for new boom to arrive. In the meantime, Mayor Don Gilman of the Kenai Peninsula Borough reported that he had purchased 1,200 feet of experimental boom manufactured at Kenai. Coast Guard Lt. Matt Carr, Capt. Rousell's representative to the MAC Group, noted that he had seen the locally produced boom and believed it suitable for light use. Carr offered 1,600 feet of heavy Coast Guard boom to meet MAC Group priorities. He cautioned that the boom was so heavy that it would have to be towed to its deployment location. For economy, the heavy boom should be placed as close as possible to its storage location in Seward. The MAC Group directed the ICT to ask Santa Maria to look at the boom and determine if it should be placed at James Lagoon or at Resurrection River.[190]

As the days went on, both the Kenai Peninsula Borough and Exxon purchased boom. The ICT's Team "F" then deployed it. The oil company continued to search North America for boom, while the borough bought and tested locally produced boom.[191]

By April 9, Team "F," using the tender Barlow and the seiners Gore Point and Katie Jean, had placed boom in Dick Creek, Middle Creek, and Nuka Bay. Finished, the team returned to Seward. At this time, Exxon had 35,000 additional feet of boom on order for Seward.[192] Comparing work done to date with as yet unprotected sensitive areas, the ICT Planning Section produced a list of "Additional Areas Recommended for Oil Spill Protection - Cape Resurrection to Nuka Island."[193]

Boom Damage and Repair

This plan for additional work turned out to be premature. On the night of April 9, a long stretch of moderately good weather ended. Fierce storms on the Kenai Peninsula coast severely damaged boom already in place. The storms produced extreme weather conditions, with 40-knot winds, 16 to 20-foot seas, and an aviation ceiling of 200 feet. The weather prevented safe aircraft or vessel operations in the northern Gulf of Alaska, along the Kenai Peninsula Coast, and at the entrance of Kachemak Bay.[194]

By April 11, the weather had moderated. Reconnaissance flights found that the storms had damaged 2,500 feet of boom deployed in Port Dick, Middle Creek, and Island Creek. The Homer Branch of the ICT launched repair attempts while the Seward ICT focused on placing 400 feet of boom on Resurrection Creek.[195]

As the weather cleared, aircraft and boat reconnaissance revealed that booms at Tonsina Point, Pederson and James Lagoons, and Delight, Middle, and Island creeks had all suffered storm

62

ADDITIONAL AREAS
RECOMMENDED FOR OIL SPILL PROTECTION
CAPE RESURRECTION TO NUKA ISLAND

Priority	Location	Amount	Resources at Risk
1	James Lagoon	1500'	wildlife
2	McCarty Lagoon	2000'	wildlife
3	Desire Creek	1500'	recreation/wildlife/waterfowl
4	Harris Bay Cr	500'	fishery/wildlife/recreation
5	Quicksand Cove	500'	wildlife/recreation/waterfowl and fishery
6	Palisade Lagoon	300'	waterfowl/wildlife/fishery
7	S. Burger Bay	1000'	wildlife/fishery

Figure 3-4

damage. The ICT developed a system for monitoring booms. It also prepared for one team to use aircraft to check the need for sorbent material at deployed booms and to verify boom failure in Port Dick and Nuka Bay. A second team was to travel by boat to Nuka Bay to repair booms there. The latter was contingent upon availability of boats. Exxon, moving in to Seward, had chartered all the boats previously used by the ICT teams.[196]

In the meantime, 2,000 feet of boom originally purchased by the City of Seward remained unused. The MAC Group reserved the boom for possible use at Nuka Island or Port Dick, or for use in multiple booming. New boom also arrived. At the MAC Group meeting on April 12, Sikstrom announced the arrival of 3,000 feet of absorbent boom. This, he said, was suitable for secondary booming.[197]

Phase-Out of Boom Operations

The MAC Group's plans for boom repair and additional booming never came to fruition. The ICT was winding down its operations, with the Coast Guard and Exxon taking over. The Coast Guard obtained copies of the boom monitoring plan and the list of boom locations. It announced that it would relocate booms that were ineffective in their original locations. The MAC Group requested that Exxon begin taking over responsibility for monitoring and maintaining boom.[198]

As the ICT's booming operation ended, the MAC Group even discussed disposal of soiled boom. The City of Seward presented plans for a plastic lined containment pit to be prepared for temporary storage until disposal methods were developed.[199]

63

SUMMARY OF BOOMING ACTIVITY
RESURRECTION BAY TO KACHEMAK BAY
April 2-12, 1989

Boom Type	Installed Size	Amount	Location	Date Installed
Curtain	36"	500'	Humpy Cove	04/02
"	"	1000'	Thumb Cove	04/02
"	"	1800'	Pederson Lagoon	04/3-4
"	"	400'	Tonsina Pt	04/05
"	"	400	Delight Cr	04/05
Sea Curtain	"	1800-3600'	James Lagoon	04/9-10
"	"	600	Port Dick Cr	04/09
"	"	1100'	Middle Cr	04/09
"	"	800'	Island Cr	04/09
Experimental Sea Curtain*	30-36"	1200'	inside Tutka Lagoon	04/09
"	24"	1100'	inside above boom	04/09
"	"	1000'	mouth of Tutka Lagoon	04/10
Curtain	"	1800'	Resurrection Cr	04/10
"	"	400'	"	04/11
"	"	100'	Seward Lagoon	04/12
"	"	1900'	Tutka Lagoon	04/12

*Experimental boom constructed by local fishermen. It was adversely affected by strong tidal currents and was placed between the mouth of Tutka Lagoon and rearing pens at the Tutka Lagoon salmon hatchery. The third boom protected the mouth of the lagoon.

Figure 3-5[200]

Booming operations in the initial phase of dealing with the Exxon Valdez oil spill had ranged from the decision to boom to disposal of soiled booms.

Comment

The ICT's staff operations made a significant difference in NPS response and the general Kenai Peninsula response to the Exxon Valdez oil spill. The team's expertise in mobilizing and dispatching resources quickly put scientists at the locations where they needed to do pre-oiling assessments. The same expertise achieved rapid booming, although boom was in short supply and the locations to be boomed were remote.

The booming operations went well despite initial concern over approval to proceed and supplies of boom. Advisors to the MAC Group developed sensible priorities for boom placement and the MAC Group adapted them to local concerns. Interagency cooperation in obtaining booms established a model of cooperation. ICT management of boom deployment was remarkably efficient, given the limited experience of almost all involved with booms. In the end, however, high energy wave action in the Gulf of Alaska swept the poisonous oil over, under, and past the booms no matter what their size. Only in sheltered waters, with multiple booms present, did the floating barriers cause the oil to hesitate.

Despite this limited success in stopping the oil, boom deployment had other values. Those values were principally psychological. Worried coastal residents saw the boom deployment vessels leaving port. They could eye ICP maps showing boom placement and have their feelings of helplessness in the face of catastrophe somewhat alleviated. This, together with other activities of the NPS sponsored ICT, was a major contribution to dealing with the Exxon Valdez oil spill.

The next phase of the incident, post-oiling collection, cleanup, and assessment, also demonstrated the adaptability of the ICS. A new element, Area Command, was added to the response mechanism to coordinate post-oiling staff and field work in Kenai Fjords and Katmai/Aniakchak.

CHAPTER 4 - COLLECTION, CLEANUP, AND ASSESSMENT

- Overview
- Area Command Operations
- Kenai Fjords Operations
- Katmai/Aniakchak Operations
- Comment

Overview

As Joe Stam, Branch Director for Homer operations of the Seward ICT observed later, the oil spill was in one way like a wildfire. The spill posed threat moved.[201] As a result, key events after the first frantic days of initial response -- continued preparations, first oilings, cleanup, and assessment -- came at different times for different points in the oil's path. Two of those points, Kenai Fjords National Park and Katmai National Park, handled response operations in different ways. They did coordinate with the Alaska Regional Office through a common mechanism, an Area Command ICT.

Area Command Operations

Regional Director Boyd Evison faced a continuing drain on his regional office staff. He asked Doug Erskine to find personnel who could set up an Area Command team to coordinate spill response activity at the regional level. Such a team, provided for in the National Incident Management System, would have the usual ICT functions. Knowing that authorities would be reluctant to release another established ICT for a non-fire incident, Evison suggested that Erskine look to NPS retirees for personnel.[202] He also appealed to other Park Service regions for temporary use of personnel. According to Evison:

> We are clearly beyond our financial and personnel resources. If we are to respond to the demands of this unprecedented emergency, we need additional assistance.[203]

Erskine struggled for more than two weeks to find suitable and available personnel. On May 11, Evison delegated his authority for oil spill response management to John Kraushaar, head of a newly established Area Command.[204] Kraushaar, a District Ranger at Sequoia/Kings Canyon National Park, became the first Area Commander. The Area Command was to:

> coordinate, support and manage those teams [ICTs for Kenai Fjords, Katmai and Aniakchak] to ensure an effective, safe and economical response to this crisis while ensuring the local managers' concerns are addressed.[205]

Evison's Line Officer's Briefing to the Area Command noted objectives. These were to provide for personnel safety, minimize impacts to cultural and natural resources, and to monitor and document effects of oil exposure. The Area Command was to support the concerns and needs of park superintendents. Evison concluded by noting that funding for the region's response to the oil spill had "not been identified."[206]

Funding had become critical. The NPS national directorate eventually recognized the problem its Alaska Region faced. It froze expenditures Servicewide. The directorate did not release the freeze until Congress authorized oil spill response spending from the NPS construction appropriation. Then when Congress, at Evison's urging, provided $7.3 million in add-on appropriations, the money went to the Fish and Wildlife Service. Some of this dribbled over to NPS.[207]

That last comment hinted at what the Area Command ended up doing. In theory, and at first in practice, the Area Command supervised activities of the ICTs at Kenai Fjords and Katmai. But this didn't last long. The area office soon became immersed in straightening out funding.

The Seward ICT and regional office had begun tracking spill response expenditures almost immediately. By mid-June it was apparent that the tracking categories in use did not meet the needs of DOI. James Randall, retired Chief of Resource Management for the Rocky Mountain Regional Office of the NPS, served as Area Command Planning Chief. He reported for duty on May 28. Randall soon found himself spending most of his time reconstructing financial records. Work to compile a report on expenditures needed for July Congressional hearings followed.[208]

Frank J. Betts, retired Superintendent of Mount McKinley National Park, arrived a week before Randall reported to Anchorage, assuming direction from John Kraushaar of Area Command. The two field ICTs looked to Betts' Area Command ICT to provide logistical support, to help with key decisions, and to serve as a link between field operations and the Regional Director. The Area Command's logistics function kept busy obtaining personnel, facilities, and supplies for the field offices. It also provided a Safety Officer who trained field personnel. Betts consulted with the field commanders on decisions about aircraft utilization, housing, and a variety of other things. He also received morning telephone updates from the field and briefed the Regional Director on this information. The Area Command was, Betts believed, a kind of mini regional office, serving the field.[209] Field activity included coordination through the ICT structure, collection and cleanup, and assessment of the oil's impact in Kenai Fjords and Katmai.

Kenai Fjords National Park Operations

Type-I ICT Phase-Out

The ICT that came to Seward the end of March left in mid-April. Pre-oiling assessment operations had ended for Kenai Fjords

68

and Lake Clark National Parks. Protective booming was in place.
With Coast Guard direction and MAC Group oversight, Exxon took over
spill response operations in Seward. Katmai National Park response
activity was directed from Kodiak. It seemed time for the Type-I
ICT commanded by Dave Liebersbach to head for home.

Kenai Fjords Superintendent Anne Castellina consented to
release the team only at the urging of Liebersbach and Park Service
regional office officials.[210] Soon after the arrival of the ICT
Bureau of Land Management officials had warned that they would need
them back at the end of three weeks. This commitment was confirmed
by Dave Liebersbach during Boyd Evison's first visit Seward in
early April. During that visit and in later meetings with other
Regional Office personnel, Evison asked that the ICT be instructed
to develop and put in place a structure for continuing response
operations. Regional officials agree, however, that the Type-I ICT
had pulled out of Seward much too quickly. Looking back, they
thought that a Type-II ICT should have been set up and operating
when the Type-I ICT demobilized. At the time, they believed that
spill response activity would taper off. To their surprise, "that
damned thing never tapered off, it just kept going and going."[211]
Suddenly, staff supporting Kenai Fjords' response to the oil spill
went from 32 on April 16, to three on April 19.[212] The three com-
prised half of Kenai Fjords' regular staff.

Transition Planning

Castellina had planned for the transition. She continued as
chair of the Seward MAC Group, which evolved to meet the changing
situation. She also devised a way to fill in after the Type-I ICT
demobilized.

Throughout the summer, Castellina continued to chair the
Seward MAC Group in addition to her other duties. The MAC Group,
continuing the daily meetings begun on April 3, set priorities and
standards for cleanup by Exxon. Amidst the false starts made after
the Type-I ICT pulled out, Castellina took a positive step with
long range benefits. Aware of the need to replicate the abilities
of the ICT Planning Section, Castellina suggested to Page Spencer
that she form a counterpart to MAC made up of agency resource
managers. The Resource MAC quickly took shape. Its members
provided expert advice on what needed to be done, for example,
priorities for beach cleanup, to the MAC Group. The MAC Group, in
turn, advised the Coast Guard and Exxon. Jack Sinclair, State
Department of Natural Resources representative to the MAC Group,
simultaneously served as chair of the Resource MAC and functioned
as a link between the two groups.[213]

Castellina and her staff anticipated that, after the ICT
pulled out, they would be able to return to their normal duties.
Simultaneously, they planned to maintain a "shadow" ICT structure
in which they would fill dual roles. Castellina would serve as
Park Superintendent and Incident Commander. Peter Fitzmaurice
would serve as Chief Ranger and Deputy ICT Commander. Bud Rice
would serve as park Resource Management Specialist and ICT

69

Operations Chief. Spencer would serve as ICT Plans Chief. A few
extra people were to be hired -- a secretary for Castellina, a
logistics person, and two public information officers. Existing
Kenai Fjords staff were to take over the ICT Finance function.[214]

The shadow ICT issued its first Incident Action Plan for the
period 0800 May 1 to 0800 May 2. The plan's objectives were:

1. Provide for the safety of all personnel.
2. Coordinate data gathering with investigators.
3. Continue resource assessment onboard M/V
   Spirit:
           assess oiled beach areas;
           assess fish and high tide habitat;
           identify dead wildlife.
4. Remove dead wildlife from park beaches.

The plan described two operations divisions. Division A, con-
sisting of Spencer and five technicians, would operate aboard
M/V Spirit. Using two inflatable boats, the team was to go from
Nuka Bay north along the coast. Division B, consisting of Vequist,
Ross Kavanagh, the Alaska Regional Office fisheries biologist, and
Stan Ponce, Chief of Water Resources for the NPS, would operate
aboard M/V Snowbird to identify dead birds from Bear Glacier south.
Simultaneously, Tort Investigator Scott Taylor was to travel by
helicopter to collect specimens from beaches in the northern end of
the park.[215]

Kenai Fjords staff managed to confirm oiling of the park's
beaches on three trips. On the first trip, an aerial observation
on April 11 before the Type-I ICT demobilized, what appeared to be
oil was sighted on the shore but not confirmed. The second, a
voyage aboard M/V Snowbird on April 13 and 14, confirmed exposure
of outer coasts and headlands to oil. Oil samples were collected
from cliffs and beaches. The third, a voyage aboard M/V Spirit,
from April 27 to May 4, accomplished an in-depth inspection of park
shoreline. Participants walked beaches looking for oil. When they
found it, they took photographs and made notes about the density,
distribution, and characteristics of the oil. Some oil samples
were collected and oil locations were marked on maps. Samples were
also collected for water quality, plankton, fish fry, and surface
soil analysis. Beaches were surveyed for bird and animal
carcasses. Carcasses were identified, counted, and arrangements
made for their retrieval. The beach surveyors collected some
carcasses to be turned over to Tort Investigators in Seward.[216]

The third assessment voyage visited 65 park beaches. Of
these, 44 were clean at the time of the visit. Eighteen were
oiled. The oil ranged from scattered splatters to saturated kelp
and debris. In many places, tar balls were melting and oozing into
sand and rocks. Sometimes surface contamination appeared only over
an area of six to 10 inches. Digging beneath the surface would
reveal a larger contaminated area, sometime extending a foot into
the substrate and spread out over a larger area. Thick oil

appeared to pool at the sand layer and flow seaward toward the water's edge.[217]

The plan for a shadow ICT didn't work. Within a few weeks of the Type-I ICT pull out on April 19, it was clear that too much needed to be done. MAC meetings had continued. Normal park summer operations were just getting started. Something had to be done about assessing oil injuries and cleaning up beaches that had been oiled. Enormous amounts of energy and time had to be devoted to dealing with the Coast Guard, with Exxon, with the press, with politicians, and so forth.

Activities directly associated with spill response, also caused other park operations to soar. Visitors to Seward flocked to the Kenai Fjords Visitors Center to get spill information. Seven-day-a-week operation, generated by spill response, coincidentally increased visitation and park utility bills. Purchasing, telephone calls, and other day-to-day business skyrocketed. Three of the park staff, Castellina, Fitzmaurice, and Rice, found themselves doing spill work full-time. This concentrated the burden of busier than normal park operations on the remaining 50 percent of park staff. Soon everyone was very tired. All felt constant stress. Dedicated to the resources they managed, the park staff also suffered emotionally as oil assaulted Kenai Fjords' pristine shoreline.[218]

At this time, Kenai Fjords spill response operations focused on two efforts. Six Coastal Rangers went to various points of the park's coastline to report oiling as it occurred. At the same time, the park boat M/V <u>Kenai Ranger</u> and M/V <u>Snowbird</u>, a contract vessel, carried scientific crews searching for newly oiled locations and reporting on the flora and fauna found there.[219]

Collection Efforts

Offshore, Exxon and the federal government deployed vessels to break up oil patches and to collect oil before it went ashore. Two U.S. Navy small craft known as skimmers, boats able to skim oil off the ocean's surface, arrived in Seward on April 8 and 9. The skimmers mounted conveyor belts of absorbent material. With one end plunged into the water, the belts removed oil from the water's surface and put it into storage tanks. The two Navy skimmers were to tow barges that could each store up to 35,000 gallons of oil; but the barges didn't reach Seward until some time after the skimmers themselves had arrived.

In the meantime, the Coast Guard planned to use its cutters <u>Planetree</u> and <u>Yocona</u> to corral floating oil with 84-inch Norwegian manufactured boom. This could begin with the arrival of a power pack necessary to inflate the boom. The first two Exxon skimmers had arrived in Seward on April 11. Two more Exxon skimmers were to arrive in Resurrection Bay the following day.

While the Navy and Exxon skimmers were getting ready, the Coast Guard cutter <u>Morgenthau</u> steamed around at the entrance to Resurrection Bay acting as mother ship to several small fishing vessels attempting to break up oil patches with herring nets.[220]

71

Local residents also formed a "mosquito fleet" of small craft whose operators simply scooped oil up in buckets from ocean waters. Additional oil collection capability was anticipated with the arrival of the Soviet ship Vaydagursky on April 15. The Soviet skimmer reportedly could skim up to 200,000 gallons of oil per hour and store up to two million gallons in its holding tanks.[221]

Even if they had been operable, the skimmers were too late to hinder the first oil from coming ashore. Storms on the night of April 10-11 began to blow oil ashore in Kenai Fjords National Park.[222] Before this the oil, which stretched from Cape Junken at the western end of Prince William Sound to Gore Point, had held offshore. The oil offshore consisted of a 20 to 32-mile-wide sheen with widely separated areas of mousse. The storms that blew the oil onshore also damaged protective booms in place at Tonsina, Pederson, and James Lagoons, Delight Creek, Middle, and Island Creeks.[223]

By April 13, overflights reported oil on several areas of the Kenai Fjords National Park coastline. U.S. Navy Skimmer No. 90, working near Nuka Bay, found the oil it skimmed too thick to pump. Five fishing boats from Kodiak were also in the Nuka Bay area attempting to break up the oil by dragging fishing nets through it.[224] By the following day, the Coast Guard had literally netted 2,000 gallons of oil but it was too thick to pump into holding tanks. The situation continued for several days and by April 15 Exxon had over 10,000 gallons of oil trapped. Available machinery could not pump the thickened and weathered oil.[225]

The inept attempts to collect oil generated some of the first local criticism of spill response efforts in Seward. Prior to this, community support had been unanimous. On April 13, local columnist Tim Moffatt observed:

> Meetings are held, maps updated, briefings given and a steady pile of paper generated. Reconnaissance flights are flown, boat trips taken, but so far, no oil has been cleaned up from the waters of Cape Resurrection or the vicinity of the bay, or from Kenai Fjords National Park.[226]

Better pumps did not arrive so the Coast Guard sought advice from a Canadian oil spill expert on how to deal with the problem. One choice was to use chemicals to thin the collected oil. This would allow it to be transferred from skimmer reservoirs, now full, to storage bladders. Environmental concerns precluded use of this method. With their reservoirs full, the skimmers were unable to pick up additional oil. A second alternative, of pumping oil within containment booms directly to storage reservoirs, awaited arrival of suitable pumps.[227]

By April 20, the Coast Guard had concentrated all of its spill response vessels operating outside Prince William Sound at Nuka Bay. North Slope crude from the Exxon Valdez had defeated the skimmers, ranging in size from the tiny U.S. Navy Marco V models to the giant Soviet Vaydagursky. Pumps proved incapable of pumping the oil directly from the water into storage tanks. Success in sucking up the weathered, carcass and debris filled oil came only with the use of two U.S. Army Corps of Engineer dredges, the Essayons and the Yaquina. Designed to dredge muck, sand, and gravel from harbors and river bottoms, the two vessels inserted their suction mechanisms under the floating oil. An efficient way had finally been found to collect the oil before it came ashore. But even these vessels had difficulty pumping the unmanageable "product" from their storage hoppers into containment barges. These problems consumed time and limited hours the dredges spent sucking up oil.

Cleanup Efforts

Realization that cleanup had to begin immediately further complicated the problem. Castellina and her staff at first believed it "ridiculous to wipe rocks and come back two weeks later and see the same beach impacted all over again."[228] They soon realized that cleanup could not be postponed until the oil had finished coming ashore. The end of injuries was not in sight. Oil washed ashore, infiltrated the substrate, washed out of the substrate back into the ocean, and then washed ashore again. Sometimes, the re-oiling occurred where the oil had originally gone ashore. At other times, the oil reinvaded the coastline at different location. It was urgent to collect the oil, whether it was afloat or ashore, to minimize the injuries it could inflict.

The National Park Service initiated the practice of assigning Resource Protection Officers (RPOs) to monitor cleanup activities. The Coast Guard recognized the value of the RPOs. At the Seward MAC Group meeting on May 8, the Coast Guard representative directed that the Park Service have RPOs present on any beach where VECO, the Exxon contractor, worked. Exxon then advised Castellina that it planned to have up to 150 workers cleaning up park beaches almost immediately. This added substantially to the Park Service's efforts to deal with the spill.

Garey Coatney, who had returned to Seward to become commander of the second ICT put in place there, estimated that a minimum of 12 RPOs would be needed at any one time to meet Coast Guard requirements. After deciding that the RPOs should be berthed on Park Service contracted boats, the ICT put three boats under contract to support the RPOs. The RPOs came from parks throughout the NPS and rotated through on 21-day assignments. This required huge efforts to manage the boats, recruit RPOs, train them, support them in the field, provide relief for rest and recreation, and provide replacements on a timely basis.[229]

Cleanup monitoring, like every other task connected with oil spill response, didn't come easily. Exxon had said that it was ready to immediately put 150 workers onto park beaches. According to Castellina "it became obvious as the summer wore on that they [Exxon and VECO] were never going to get it together enough to get

Beach cleanup crew.   (Photo courtesy
of Karen Jettmar.)

Steam-cleaning oiled rocks.   (Photo
courtesy of Karen Jettmar.)

these people on Park beaches. . .there was actually a period of over one month that were was no cleanup activity in the Park at all--none."[230]

Exxon/VECO undertook two types of beach cleanup. "Type A" consisted of surficial cleanup in which crews, working without tools, picked up and removed oiled debris from a beach. "Type B" consisted of surface and subsurface cleanup in which crews, working with shovels, excavated oiled materials from a beach. Alternatively, low or high-pressure hoses washed oil from rocks and boulders on beaches. "Bioremediation" was another cleanup technique. Biological warfare against the oil, bioremediation meant spreading chemicals that nurtured microbes on a beach. The multiplying microbes were then expected to eat the oil. Standing apart from other agencies, the Park Service banned bioremediation as a cleanup technique to be used on park lands. They didn't know enough, said park officers, about the long-term effect of bio-remediation and other chemical treatment. The formula for the chemical compound, known as Inipol, is a closely guarded secret, but Inipol is known to be toxic to marine life.[231]

About five percent or 20 miles of Kenai Fjords beaches received oil. Park officials authorized Type A cleanup for all of them. In some cases, delay in starting Type A activity meant that beaches deteriorated in situations where Type B cleanup needed to be undertaken.[232]

Quartz Bay, about 40 miles southwest of Homer, at first needed only Type A cleanup. By the time cleanup crews reached Quartz Bay, oil on the beach had melted and sunk below the rocks. It became a Type B beach. Reluctant to authorize a Type B cleanup, park officials told Exxon to bypass the beach. It was never cleaned up. Other beaches in the park receiving Type A cleanup included Beauty Bay, Pony Cove, Bear Glacier, Porcupine Cove, Noname Cove (south of Porcupine), and Agnes Bay. Aialik Bay, where crews shoveled up oil-soaked rocks and soil; and Black Bay, Taroka Arm, and Verdant Cove, where crews hosed down rocks with hot water washes, received Type B cleanup.[233]

Kenai Fjords Incident Command Team

Normal park operations spurted because of oil spill activity. This, plus coordination of post-oiling investigations and oversight of cleanup work quickly made Castellina realize that Kenai Fjords needed additional help. The weeks that followed demobilization of the Type-I ICT made it apparent that the shadow ICT demanded too much of her park staff. As a result, she requested a new and smaller ICT. It opened for business in Seward on May 16. The ICT staff included 14 NPS and two Forest Service personnel.

The new team rented office space across the street from the Kenai Fjords Visitors Center. It found the waters off Seward swarming with marine traffic. Fifty-three boats were supporting oil booming, skimming, and so forth. Fourteen more were on standby. Dispatchers daily sent another 18 boats to pick up birds.

75

Twenty-three small boats (the mosquito fleet) picked up oil. Two other boats supported beach cleanup.[234]

The first Incident Action Plan issued by the Kenai Fjords ICT listed four strategic objectives:

1. Maintain personnel safety as the highest objective.
2. Continue to protect environmentally sensitive areas.
3. Cleanup and minimize further oil deposition on beaches at the upper end of Resurrection Bay.
4. Remove free oil off shore and in Nuka Bay.[235]

This initial attempt to assume the broad responsibilities of the Type-I ICT, that is to work outside park boundaries, quickly shrank to a more limited concern. Similarly, an initial attempt to relieve the park superintendent of oil spill responsibilities also resolved itself. In the traditional ICT structure it is always clear, and stressed, that the ICT serves the line manager for the land manager. After a period of adjustment, this became the circumstance under which the Kenai Fjords ICT worked.[236]

Once Kenai Fjords ICT was in place, long dreary days of stress followed. Sometimes visiting dignitaries broke up the drudgery of dispatching RPOs and coordinating cleanup efforts. On June 1, Coast Guard Commandant Adm. Paul A. Yost, accompanied by Federal On-Scene Coordinator Vice Adm. Clyde E. Robbins, visited Seward.[237]

Costs for the drudgery ran high. Weekly costs included $18,000 for the RPOs, $14,000 for the ICT's overhead staff, $9,000 for the Tort Team, and $2,000 for the Coastal Rangers. In additional to individual transportation costs to move people to and from Seward, weekly in-park transportation costs included $62,000 for three vessels and $10,000 to $25,000 for air support. Added to these figures were expenses for supplies, lodging, and ground transportation at Seward.[238]

By July 1, response operations at Seward had stabilized with eight overhead personnel and 9 scientists with 18 Coastal Rangers and RPOs in the field. The focus of operations had shifted only slightly from earlier objectives, with concentration on cleanup activities. Safety remained paramount and safety concerns expanded to include contract personnel. Park resources were to be protected from unacceptable effects resulting from cleanup. Spill impacts and treatment areas were to be identified. Research and cooperative activities were to be supported as requested.[239]

Twelve days into July, Glen McCrory, Exxon Incident Commander at Seward, expressed surprise when told that no cleanup was underway in Kenai Fjords National Park. The RPOs, sent into the field and maintained at great expense, had nothing to do. When McCrory asked if Exxon could pay for the RPOs, he had to be told that the NPS had no way to accept private funds for that purpose and, in fact, had been directed not to find a way.[240]

By the end of July, the Kenai Fjords operation was down to four overhead personnel, three scientists, and six Coastal Rangers and RPOs.[241] Besides monitoring ongoing activity, the ICT looked ahead to tasks that would have to be faced during the winter, spring, and summer of 1990.

In reviewing the incident, Castellina described Kenai Fjords' oil spill response as having three phases. Phase I began March 24 when the Exxon Valdez grounded. During that phase, the park worked to document its coastal resources as they existed before any oil reached the park's shore. Phase II began on April 10 when oil from ruptured tanks of the Exxon Valdez hit Kenai Fjords. During that phase, the park worked to continue documentation of its resources, assess the impact of oil strikes, and monitor shoreline cleanup. Phase III would begin when Exxon shut down its 1989 cleanup operations. During that phase, the park would work to continue its resource studies and to monitor oil already on the shorelines or to be purged from Prince William Sound.[242]

Katmai National Park/Aniakchak National Monument Operations

Katmai National Park and Aniakchak National Monument were the hardest hit of any NPS areas. Collection, cleanup, and assessment developed much differently for these areas than they did for Kenai National Park. Geography caused some of the difference. Headquarters for Kenai Fjords National Park were in Seward. This was only minutes by air and a few hours by boat from the scene of spill collection, cleanup, and assessment activity. Headquarters offices for Katmai and Aniakchak, were in King Salmon, an air hub in the interior of the Alaska Peninsula. Convenient to interior portions of Katmai, King Salmon is separated from the park's coastline by 80 to 100 miles of rugged landscape that includes a mountain range. The coast can be reached from park headquarters only by light aircraft able to land on beaches or lagoons.

As in Kenai Fjords National Park and Preserve, spill related activities took their toll on NPS staff in Katmai, Aniakchak, and Kodiak. Bane, Meldrum, Hamson, Roy, and Blinn all worked long hours and were continually on call. They devoted tremendous amounts of time to dealing with the Coast Guard, Exxon, the media, politicians, and Alaska Regional Office. All employees in Katmai and Aniakchak were personally impacted by the assault on resources they managed and the need to maintain normal park operations in the face of unprecedented activity. The resources of these already minimally staffed areas were drained even further by oil spill response demands.[243]

Preparation for Oiling

Four days after the oil hit Kenai Fjords National Park, aerial reconnaissance spotted a 40-mile slick of mousse and sheen moving down Shelikof Strait between Kodiak Island and the Katmai coastline. Observers spotted light oiling on Cape Douglas on April 12.[244] On April 18, additional oil was spotted at Kukak Bay, 50 miles south of Cape Douglas. Later in April, spotters confirmed the presence of oil further south at Missak and Kashvik Bays.

Ray Bane, Superintendent of Katmai
National Park and Preserve, notes
mousse on beach. (NPS photo.)

Ray Bane dips mousse from tide pool
on Shaw Island. (NPS photo.)

Ray Bane and oiled bird on
beach in Katmai National Park
and Preserve. (NPS photo.)

Subsequent sightings identified oil at Hallo Bay and other locations on the Katmai coast.[245] Nancy Deschu's pre-oiling assessment team observed only small patches of oil on Cape Douglas the morning of April 26. After a storm that same day, Deschu documented Katmai's first major oil strike on Cape Douglas.[246]

Katmai had begun preparing for the onslaught of oil early in April with the assistance of Dan Hamson and Cordell Roy from the regional office. After pre-oiling assessment surveys were launched from Homer in mid-April, both flew to Kodiak where they joined Superintendent Ray Bane who had been in town for two days. They found Kodiak's emergency response structure operating out of Kodiak Borough offices. Borough officials welcomed them and provided unlimited use of telephones, copying machines, and other support since the NPS had no facilities in Kodiak.[247]

Roy, serving as Ray Bane's representative at Kodiak, encountered the Type-II ICT sent to Kodiak to work for the Fish and Wildlife Service. Since Fish and Wildlife wasn't using the team, Roy put it to work organizing the Katmai/Aniakchak oil spill response. The Kodiak ICT, sometimes called the Kodiak or Katmai Field Office, issued its first Incident Action Plan on April 16. The team began operating with 15 people from the Bureau of Land Management, and one each from the Fish and Wildlife Service, NPS, and University of Alaska, Fairbanks. By the following day the Kodiak ICT picked up scientists and vessels at sea sent from Homer to do pre-oiling investigations of the Katmai coastline. This brought the total of Park Service personnel accounted for by the team up to 24.[248]

At first, in addition to wrapping up the pre-oiling assessments launched from Homer, Katmai work undertaken at Kodiak focused on identifying areas for feasibility and priority of booming. Planners also prepared to form a post-oiling assessment team. Traveling with Fred Brew, an Exxon contracted booming expert, Bane flew the Katmai coast to identify areas suitable for booming. Later Roy arranged for the same expert to fly the Aniakchak coast. Roy developed information from these surveys into a booming priority list that he presented to the Kodiak emergency management council. With characteristic wide open bays influenced by tides and currents, Katmai and Aniakchak had few areas suitable for curtain booming. There was some hope that deflection booming, in which booms could be deployed to deflect oncoming oil from entering the bays, might work.[249]

Collection Efforts

Boats with curtain and deflection boom were in route between Kodiak and Katmai when massive amounts of oil poured through Shelikof Strait and hit the Katmai coastline, consequently, no boom was in place when the oil struck. The Russian skimmer Vaydagursky, the largest such vessel in the world, looked from the air like a waterbug attempting to chew up a mass of oil flowing all around it. The Russian ship did manage to skim up 63,000 gallons of oil between April 30 and May 1, but vast quantities of deadly petroleum product remained. The horror of the situation was highlighted when the skimmers began to encounter many live, oiled, birds trapped in the mousse. On May 1, winds blowing out of the southwest began pushing oil ashore on beaches not previously contaminated. By May 2, 15 vessels including the Soviet ship, two U.S. Army Corps of Engineer dredges, and nine skimmer boats were working off the Katmai coast.[250]

None were any more successful in dealing with the oil than the skimmers operating off Kenai Fjords coastline. Coast Guard authorities had ordered the boats with boom enroute to Katmai to stay with the oil, so little boom was placed. Some boom went out in Hallo Bay, but the lack of suitable sites and the arrival of the oil precluded much booming at Katmai.[251]

The untimely winds, lack of booming, and masses of oil combined to injure Katmai's coastline. Flying by helicopter on May 2 from Kashvik Bay at the southwest end of the park to Cape Douglas, Bane counted six dead birds per 100 feet over a six-mile stretch of the Hallo Bay beach. On the same flight, Bane saw 14 bears feeding on dead, oiled carcasses.[252]

Cleanup Efforts

With the opportunity for booming past, Roy and the ICT at Kodiak turned their attention to cleanup. As at Kenai Fjords, cleanup was slow in starting. Somehow concern about the slow start reached Washington. About 2200 on May 3 Coast Guard authorities in Kodiak received a call from the White House. Coast Guard Vice Adm. Robbins, now overall Federal On-Scene Coordinator for the oil spill, was headed for Kodiak. He wanted to see oiled beaches in Katmai National Park and the Becharof National Wildlife Refuge. At 1000 the following day, Adm. Robbins and Roy helicoptered to Katmai's coastline. Visibly affected by the devastation of a beach that smelled like a refinery with oil rolling in the surf and smearing the sand, Adm. Robbins turned angrily to Roy. "Why," he said, "are you [the NPS] obstructing us? We could do a lot of good here with cleanup crews." Roy advised the admiral that NPS had approved beach cleanup some days previously. Evison had, in fact, given approval the same day it was requested. Told that no cleanup crews had since appeared, Robbins radioed ahead for Exxon officials to meet him when the helicopter returned to Kodiak.[253]

Exxon soon had one crew of 50 people removing oiled debris from Katmai's beaches. By early May the oil had contaminated most of Katmai's hundreds of miles of coastline. The first crew went to Cape Chiniak at the entrance to Hallo Bay. It worked for the better part of a month shoveling oil contaminated materials into bags for removal from the beach. Additional crews followed to pick up oiled debris and carcasses from other parts of the Katmai coastline. Similar work followed for Aniakchak, hit by oil on July 21, although storms prevented verification that oil had hit the monument's beaches until after July 4.[254]

Bane decided to require that Resource Protection Officers accompany cleanup crews at Katmai and Aniakchak because of the large number of bears that would be encountered. The Coast Guard then imposed this requirement for crews working on any NPS lands, which meant that RPOs had also to be sent to Kenai Fjords.[255]

Storms and confusion caused the delay in getting Katmai clean-up operations fully underway. After the initial crew went to Cape Chiniak, storms drove the crews into Kukak Bay for several days. Then Exxon sent all crews to the Kodiak Island coastline across Shelikof Strait for about ten days. Complaints to the Coast Guard and Exxon brought cleanup crews back to Katmai after Exxon had hired more crews. Even then, special effort was required to assure that the cleanup crews followed Park Service priorities. In one instance, Gilbert B. Blinn, a former Katmai Superintendent who replaced Roy as Superintendent's Representative at Kodiak on May 16, received a radio message from one of the boats working off Katmai. Blinn learned that Exxon had ordered the cleanup crews to Katmai Bay the following day. Since cleanup of Katmai Bay was a lower priority, Blinn asked Exxon to order the boats to remain at Cape Chiniak. Exxon refused. Only Coast Guard intervention forced Exxon to honor NPS priorities.[256]

It was not until mid-July that cleanup was in full swing for Katmai and Aniakchak, with three crews each supported by seven or eight vessels at work. Accompanied by RPOs to protect them from bears and the bears from them, workers would walk line abreast down beaches picking up debris and placing it in bags. All-terrain vehicles (ATVs) would then move accumulated bags to the beach where they were transported by small boat to larger vessels standing off shore.[257]

Cleanup efforts were costly, both in dollars and in environmental impact. Exxon, working through its subcontractor VECO, spent about $200,000 per day on efforts to cleanup the Katmai coastline. At Cape Chiniak, 40 workers spent about three weeks removing 200 tons of oiled material, yet none of the area was free of oil when the workers quit. Cleanup operations added to the distress caused by the poisonous oil. Low flying, fixed-wing aircraft and helicopters alarmed nesting birds. Bear sows with young cubs exhibited harassment responses to the unprecedented human activity in their usually solitary habitat. In one case, at Kukak Bay on May 19, a VECO employed "bear guard" killed a brown bear when it threatened workers.[258] These impacts vindicated Evison's statement to a Senate appropriations subcommittee, shortly after the spill but before the first strikes on parklands, that care must be taken that cleanup work did not become more damaging than the oil itself.

Cleanup crews picked up incredible amounts of oil-soaked debris from the park's beaches. On June 24, 55 people working from the M/V Ocean Tempest picked up 1,789 bags of "spoil" from 201 yards of beach at Kaflia Bay. The next day they found a new four-foot wide band of oil on the beach and picked up 2,913 more bags of spoil.[259]

By July 27, four areas of the Katmai coastline had been subjected to cleanup work, or "treated." These were Cape Chiniak and Chiniak Lagoon, Hallo Bay beach and lagoon, the south shore of Cape

81

Gull and Kaflia Bay, and Cape Douglas. Beach assessors identified eighteen areas, including Cape Chiniak, Cape Douglas, and Hallo Bay, as needing initial or additional cleanup. Park officials recommended only Type-A cleanup.[260] By early August, crews had removed some 56,000 bags of spoil from 65,000 yards of Katmai beaches and 66 bags of spoil from 16 miles of Aniakchak beaches. By mid-August at Katmai, another 18,000 additional bags of spoil had been filled and 23,000 yards of coastline covered. On September 15, after cleanup efforts stopped, the total count was 95,151 bags of spoil collected from 111,585 yards of Katmai coastline and 154 bags of spoil collected from 35 miles of Aniakchak coastline. Biologists counted over 8,400 dead birds along Park Service shoreline. Overall, it appeared that approximately 320 of Katmai's 398 miles of coastline had received oil, as had about two-thirds of Aniakchak's 68 miles of coastline.[261]

Post-Oiling Assessment

While collection and cleanup efforts were underway, the Katmai and Aniakchak staff and ICT started assessing oil effects. Janis Meldrum went to Kodiak after briefing the investigators sent out from Homer in mid-April. At Kodiak, Meldrum served as ICT Operations Chief. She formed three or four-person resource crews that went to Katmai beaches to fill out assessment forms. On their return to ICT headquarters, crew members updated maps of spill impacts. This information guided cleanup priorities.[262] By June 7, the resource crew had conducted 80 beach assessments. They included information on beach substrate, degree of oiling by tidal zone, and photo documentation. Because oil mixed with substrate was very difficult to see from aircraft, the on-scene work of the resource crew also helped to describe movement of the oil.[263]

While the resource crew activity continued throughout the summer, Katmai also brought in bio-technicians to do more comprehensive resource surveys. Work focused on gathering information to support the damage assessment process under the Comprehensive Environmental Response compensation and Liability Act (CERCLA). Long term transects were begun by a variety of agencies and private contractors for determining the fate and persistence of oil. The studies were mostly funded by Exxon.[264]

Will Troyer, retired Wildlife Research Biologist in the Alaska Regional Office, headed the bio-technicians who reported for duty on June 19.[265] The new arrivals received four full days of training. Training subjects included orientation to Katmai and the oil spill, resource monitoring, beach assessment, data collection and storage, park regulations, field notes and paperwork, photography, equipment, documentation, boating safety, and seamanship.[266]

Although Bane and his staff considered stationing the bio-technicians at key points along Katmai's coastline, anticipated communications and transportation problems led to a decision to support them from a vessel. The ICT selected a 70-foot yacht, the Staccato, for the bio-technicians. The vessel, at a cost of $3,500 per day, provided staterooms for two to six researchers, computer,

office space, and inflatable boats for beach access.[267] One bio-
technician crew of two did go into Aniakchak National Monument and
worked from a cabin at Aniakchak Bay. The Olympic National Park
team that did the pre-oiling assessment returned and also did
post-oiling investigation for the monument.[268]

The resource monitoring crew found that the permanent fishery,
vegetation, and intertidal plots established before the arrival of
oil were, for the most part, in locations not affected by oil.
While these were useful as controls, it became necessary to
establish new plots in both lightly and heavily oiled areas. Plans
were made to monitor the plots for several years.[269]

The crews also found that oil spill injury was continuing. At
Shakun Islands, seven miles northeast of Cape Chiniak, no young
gulls were present. Pools of oil were on the beaches and boulders.
Oil continuously seeped into the sea where sea otters were
swimming. Biologists estimated that the 7,000-plus carcasses of
dead, oiled birds recovered were only about 25 percent of the
killed birds.[270]

The bio-technicians also continued beach assessments. By
August 12, the shore-based Aniakchak crew, the _Staccato_-based crew,
and a crew operating with a Bell 206 helicopter had completed over
200 assessments on the Katmai coast and 40 on the Aniakchak
coast.[271]

Besides establishing new plots and doing beach assessments,
the bio-technicians collected data for various wildlife populations
and documented the productivity of various colonies. They also
looked for contaminated birds, eggs, nesting material, and egg
shells. The biologists assessed the status of sea lions, hair
seals, and sea otters occupying rocks within Katmai National Park.
They placed emphasis on haul-out areas affected by oil and on oiled
animals. The teams also surveyed fox dens and scat to determine if
small mammals feeding on beach carrion were affected by the oil.[272]
Thirty bears were collared on the Katmai coast to determine the
effects of oil on them.[273]

In making a preliminary report on the bio-technicians'
findings during the summer, Troyer noted that over 300 miles of
Katmai's beaches had received some oil impact. Many miles were
heavily oiled. He said that if a significant amount of oil
remained in 1990 and additional cleanup were undertaken, it would
be necessary to repeat detailed beach assessment similar to that
done in 1989. There should, he wrote, also be repetition of the
1989 ecological, bird, and mammal surveys. Intertidal and vegetation transects and water quality stations required monitoring.
Seabird, raptor, marine mammal, brown bear, and archeological
surveys would be needed.[274]

Comment

Troyer's recommendations reflected the reality that the _Exxon
Valdez_ oil spill was not ending with the summer of 1989. Oil from
the gigantic vessel's ruptured tanks remained on the shores of
Kenai Fjords National Park, Katmai National Park and Preserve, and

Aniakchak National Monument. The field ICTs and the Area Command ICT demobilized in mid-September. Continuing oil spill concerns led the Alaska Regional Office to establish a separate Office for Oil Spill Response. Nonetheless, the initial NPS response to the oil spill had ended.

The final phase of the initial NPS response to the oil spill differed in significant ways from earlier phases of that response. One key difference was that NPS was a peripheral, rather than principal, player in the collection and cleanup effort. While Park Service officials set priorities for and approved cleanup efforts on park lands, the actual cleanup was done by a third party. This weakened NPS control of the cleanup, even in instances where RPOs accompanied cleanup crews. In Katmai NPS resource personnel had to compete for time on ICT contract helicopters.

A second key difference was the attempt to impose an Area Command structure on the ICT operation, and to staff that structure with NPS retirees. This try at asserting line authority through what functioned as a staff activity in the regional office quickly ran afoul of the great authority NPS gives to its superintendents.

Rotation of the office and field personnel every twenty-one days posed significant management problems. Resource managers called in to assist in oil impact assessment, incident commanders, resource protection officers, and administrative staff were all hired for twenty-one day periods. This frequent rotation of personnel created operational inconsistencies. As a result, the coordination of collection, cleanup, and assessment suffered. Each incident commander brought with him a new perspective on park operations and priorities. Resource management specialists each had their own ideas of how assessment should be handled. Specific examples of this include differences in the way RPOs were handled and in a continuing lack of standardization of data collection techniques.

Despite these problems, the area command and field ICTs continued to be useful tools in responding to the oil spill incident. How effectively those tools were used, and the story of the dedication and determination of those who wielded them will become better understood as the passage of time increases perspective about the response to the oil spill.

CHAPTER 5 - PERSPECTIVES

Overview

The ICTs demobilized. The last Exxon-paid cleanup workers left. A few volunteers remained at work after the oil company shut down operations on September 15, 1989. This reflected public dissatisfaction with Exxon's cleanup efforts, as did a State of Alaska announcement that it would make funds available for continued cleanup work throughout the winter.

This historical perspective was begun shortly after the Exxon Valdez oil spill occurred. Although the events surrounding the oil spill and the NPS response may be later reinterpreted by historians with new perspectives, contemporary historical examinations, such as this report, can immediately serve as management tools. Such historical narratives, combined with other procedures, can help an organization prepare for future challenges. It is with the latter purpose in mind that this chapter highlights some of the conclusions of previous chapters.

Background and Initial Response

Looking back, it is clear that NPS should have been much more aware of the threat posed to its coastline by a major oil spill near the Trans Alaska Pipeline marine terminal at Valdez. Cargoes of millions of gallons of North Slope crude oil daily left the terminal for transits through some of the roughest waters in the world. Some 8,700 such sailings had taken place between the time the marine terminal at Valdez went into operation and the time the Exxon Valdez ran aground. In addition to vessels heading south, other tankers steamed southwest along Kenai Fjords National Park's coastline to enter Cook Inlet and deliver oil to refineries on Cook Inlet.

Oceanography and history both predicted where the oil would go if spilled. The Alaska Coastal Current was a well-known entity. The small 1970 spill, its existence buried in the files of Katmai National Park and Preserve and in the memories of retired NPS personnel, had carried oil to the shores of the Kenai and Alaska peninsulas.

Boyd Evison knew that an oil spill was one of many ways in which park resources might be adversely affected by accidents or routine modern human activity. For over three years prior to the spill, Evison had worked to develop and implement a Region-wide

Science Initiative. In place, the program would have greatly reduced costs and increased effectiveness of preparation for and response to the spill. Although Evison anticipated the possibility of a spill and attempted to implement a program in preparation, the threat was not widely acknowledged. Even if the National Park Service had recognized the threat, it was not prepared to deal with the unprecedented onslaught of 10.8 million gallons of oil. Prior and firm identification of the threat would have made easier NPS's task of convincing sister agencies, DOI, and others that its coastlines were at risk.

Instead, awareness came in several ways. Dave Ames knew by intuition and reasoning. Anne Castellina knew because of Bud Rice's graduate study and Tom Royer's map brought by Senator Stevens' on his visit. Dan Hamson and Cordell Roy knew as a result of Hamson's quick foray to the library. Each arrived at the same conclusion. Oil flowing from the ruptured tanks of the Exxon Valdez threatened Kenai Fjords National Park, Katmai National Park and Preserve, and Aniakchak National Monument. Greater certainty of that risk might have intensified efforts by DOI staff and NPS officials to assure that NPS had representation in the DOI Coordination Center at Valdez. Such representation would have put NPS in "on the ground floor," as one oil spill participant later observed. As it was, officials in Valdez focused on Prince William Sound, where oil was ankle-deep on some island beaches. They seemed to regard NPS concerns as alarmist. The Service, after all, was clamoring about coastlines hundreds of miles away. Their opinions certainly influenced departmental and other officials. This view from Valdez complicated the Alaska Region's effort to get the resources necessary to prepare for the oncoming oil.

As soon as the oil-posed threat was clear, within days of the Exxon Valdez oil spill, involved NPS personnel recognized two problems. The first problem was that the National Park Service lacked information about the threatened resources. The second problem was that the National Park Service needed a mechanism to correct the first problem and to deal with other effects of the rapidly approaching oil.

Spencer and Rice, returning from their overflight of Prince William sound, recognized the first problem and drafted a solution. Evison and Ames, in conversation before arrival of the first ICT, came to the same conclusion. Castellina, in speaking with Ames, recognized the need for a mechanism to deal with the threat. Ames and his regional staff knew of and obtained, a satisfactory mechanism, the Incident Command System. The Incident Command System with its quasi-military staff structure provided a means for quickly organizing a response to the oil spill threat. Calling in the ICT paralleled and anticipated what President Bush did on April 6 when he called in the military to take advantage of its organizational and logistical expertise.

NPS was, and is, fortunate to have employees such as Rice and Spencer with initiative and resourcefulness. It is also fortunate to have courageous managers such as Ames, Bane, Castellina, Evison, and Galvin. Not all officials would have defended the resources for which they were responsible in the face of high-level opposition. The tasks of the managers would have been easier had they had advance knowledge of the requirements of the Clean Water Act and CERCLA. They were fortunate to have Roy and Hamson available, who did have such knowledge. The pre-oiling assessments, remarkable for their intuitive implementation, could have benefited from knowledge of pollution injury assessment evidence standards. The managers' constant struggles with Coast Guard and other officials might have gone more smoothly with prior understanding of the rules that govern pollution responses. Despite this, the initial response of the NPS to the Exxon Valdez oil spill speaks well for the abilities and determination of its personnel at the national, regional, and park levels.

Command, Control, and Coordination

What followed initial response to the Exxon Valdez oil spill also speaks well for the National Incident Command System and the Type-I Alaska ICT. The personal histories of Dave Liebersbach and his core team, particularly their work on the Yellowstone fires of 1988, gave them appropriate experience to deal with the complex technical and political problems they faced on arrival at Kenai Fjords National Park. Their personal qualities--Liebersbach's command presence, his core staff's expertise--gave them the ability to use that experience effectively.

Liebersbach's recommendation to Castellina that she form a MAC Group smoothed the way for coordination of conflicting needs and priorities. Castellina's leadership of the group proved particularly effective, as did the unique contributions of Kenai Peninsula Borough Mayor Don Gilman. Only the attempt to extend the MAC Group's authority to Homer operations flawed its record. A combination of political circumstances and transportation and communications difficulties doomed that attempt. Castellina's first instinct, to set up a separate MAC Group in Homer, should have prevailed even though not in accord with Incident Command System policies. The eventual real independence of the Homer advisory committee resulted in a de facto independent MAC Group there. This came only after initial, time consuming efforts to establish control by the Seward MAC Group. Those efforts clogged and confused deliberations of the Seward MAC Group when it needed to be making rapid and informed decisions about operations in the Seward area. The initial confusion over the Homer committee's role had similar impact in Homer on activities of the advisory committee there.

The requests of Katmai and Lake Clark National Parks and Preserves for ICT assistance were timely and well justified. ICT Branch operations from Kenai served Lake Clark quickly and well. ICT Branch operations from Homer were less effective for Katmai because of the distances involved and rough waters separating Homer from the Katmai coastline.

NPS staff at Katmai National Park and Aniakchak National Monument initially met resistance to their requests that oil-spill resources be expanded to include the coastline of those parks. Distance and lack of coastline accessibility made it difficult for Katmai staff to establish credibility of their concerns. This hesitation resulted in serious delays in pre-oiling assessment and boom deployment. Establishing operations headquarters for Katmai and Aniakchak in Kodiak improved coordination of collection, clean-up, and post-oiling assessment activities.

Still the job got done. The drawbacks of using Homer as a base, done mostly to accommodate the needs of Gilman, may have been outweighed by his invaluable service as a NPS ally.

Pre-Oiling Staff and Field Operations

Day-to-day staff and field operations in the pre-oiling assessment phase of NPS response went smoothly. Liebersbach's highly skilled Type-I ICT made the difficult look easy. Great experience and many 18-hour days lay behind the efficient and calm demeanor of his ICT members.

The early decision to call in a NPS Civil Litigation, or Tort, Team helped to establish chain of custody for most evidence collected by the pre-oiling investigations. Greater familiarity with CERCLA and the Clean Water Act and their standards for evidence would have been beneficial. As it was, attempts to obtain copies of CERCLA and Clean Water Act standards were not successful until much pre-oiling assessment work had already been done. Data collection also suffered to some extent from two other factors. The first of these, occurring within each park's assessment effort, was the tendency of individual scientists to modify techniques according to their own experiences. The second, occurring in coordination of evidence collection between parks, was the tendency of ICT resource advisors themselves to modify techniques according to their own experiences. Thus data, which ideally should have been collected and expressed in standard fashion, were somewhat different according to by whom and where they had been gathered. Despite these problems, the pre-oiling assessment was a tremendous challenge, well met. Initiated at the beginning of April, it finished just in time, before oil struck park coastlines. For Kenai Fjords this was mid-April and for Katmai and Aniakchak, late April.

Booming operations, conducted concurrently with pre-oiling assessment, were controversial at the time. Coast Guard officials thought the boom would not be effective given the magnitude of the spill and the high energy wave action along park coastlines. Efforts to obtain boom, establish priorities for its use, and place it required considerable efforts by the MAC Group and by the ICT. Little information is available on how effective the boom was in repelling oil. It is likely, however, that refusing to deal with booming would have materially degraded effectiveness of the ICT. Consequences that might have ensued could have included competition for resources such as boats and crews, disintegration of the MAC Group, and a lasting enmity toward NPS.

Collection, Cleanup, and Assessment

The collection, cleanup, and assessment phase of NPS response to the Exxon Valdez oil spill involved a major change in the way in which NPS related to on-going activity. In the pre-oiling phase, NPS not only directed activity, but was the principal active participant. Its scientists, for the most part, did the assessment work while the ICT facilitated that work. In the post-oiling phase, the Coast Guard directed Exxon contractors as they collected oiled materials and cleaned shoreline. The National Park Service, using an Area Command and Field ICTs, provided advice and priorities for cleanup work. The same ICT structure also supported NPS field personnel. Field personnel included Resource Protection Officers, or "bear guards," accompanying cleanup and collection crews. Post-oiling scientific assessment teams were also in the field.

Four key decisions characterize the post-oiling phase of the National Park Service response to the Exxon Valdez oil spill. These were: (1) to maintain the ICT structure and do so with the assistance of retired NPS personnel; (2) to establish an Area Command ICT; (3) to permit shoreline cleanup by Exxon contractors; (4) to have Resource Protection Officers accompany contractor's collection and cleanup crews.

Kenai Fjords' brief experience at attempting to manage post-oiling oil spill response without ICT help demonstrated the need for that assistance. The regional office's decision to call upon an Area Command ICT for help was also amply justified. Post-oiling ICT use did quickly run into problems.

NPS gives substantial authority to its individual park superintendents. Superintendents, responsible for all aspects of operations and protection of resources within their units, report to regional directors who report to the Director of the NPS. NPS staff offices advise, but do not control, superintendents. The Incident Command System requires that land managers delegate their authority, insofar as it pertains to a particular incident, to Incident Commanders. In most situations this works well, for the ICTs function as line agents, direct extensions of land managers' authority.

In the post-oiling phase of NPS response to the Exxon Valdez oil spill, the two field ICTs functioned well. There were some misunderstandings as highly trained ICT personnel left and less experienced personnel filled the vacancies they left. Throughout the second phase of the incident, rapid turnover of personnel proved to be costly in terms of effectiveness and travel expenses. The Area ICT seems to have functioned, not less well, but perhaps in a different way. Originally conceived of as overseer of the field ICTs, the area command soon evolved into a staff activity of the Alaska Regional Office. The Regional Director discussed the Area Command ICT's role with his superintendents before establishing it, but Incident Command System and NPS organizational philosophies clashed immediately. These relationships were further complicated by use of NPS retirees, imbued with the NPS philosophy, for Area Command staff. Predisposed to accept superintendents' assertion of their prerogatives, the retirees gravitated to a staff versus a line role. The result was an Area Command ICT that exercised less initiative than anticipated. While not critical, this probably exacerbated the lack of standardization of field work that first occurred in the pre-oiling assessment phase of spill response.

That pre-oiling assessment lack of standardization may be attributed in part to insufficient knowledge of pollution laws and regulations, and of the way oil spills behave. The same unfamiliarity with oil spills probably prompted the hesitation to authorize immediate cleanup. As events evolved, the hesitation quickly gave way and had no practical effect.

High levels of bear activity on the Katmai coast led to the decision to require that Resource Protection Officers accompany collection and cleanup crews working there. Transferred without amendment by the Coast Guard to Kenai Fjords, the requirement was probably less necessary there. It did provide for NPS monitoring of the crews, which was desirable.

Conclusion

The Exxon Valdez oil spill demanded, and got, the best efforts of those involved. NPS employees in Katmai, Aniakchak, and Kenai Fjords were called upon to maintain park operations with reduced resources in already minimally staffed areas with increased visitation. Those directly involved in oil spill related activities were constantly on call by the media, politicans and Alaska Regional Office, resulting in excessively long working hours. Employees in the parks, as well as those brought in from other areas to aid in the oil spill response were under constant stress. The aggressive performance of ICT members, the concern for park resources showed by municipal officials and residents of the Kenai Peninsula and Kodiak, and the courage and determination of NPS stewards of cultural and natural resources made the spill response successful. Perspectives on execution of the response highlight some things that might have been done differently. They do not detract from the achievements attained. NPS is fortunate that the awful circumstances that created the Exxon Valdez oil spill were concurrent with the circumstances that made exceptional people available to respond to the spill.

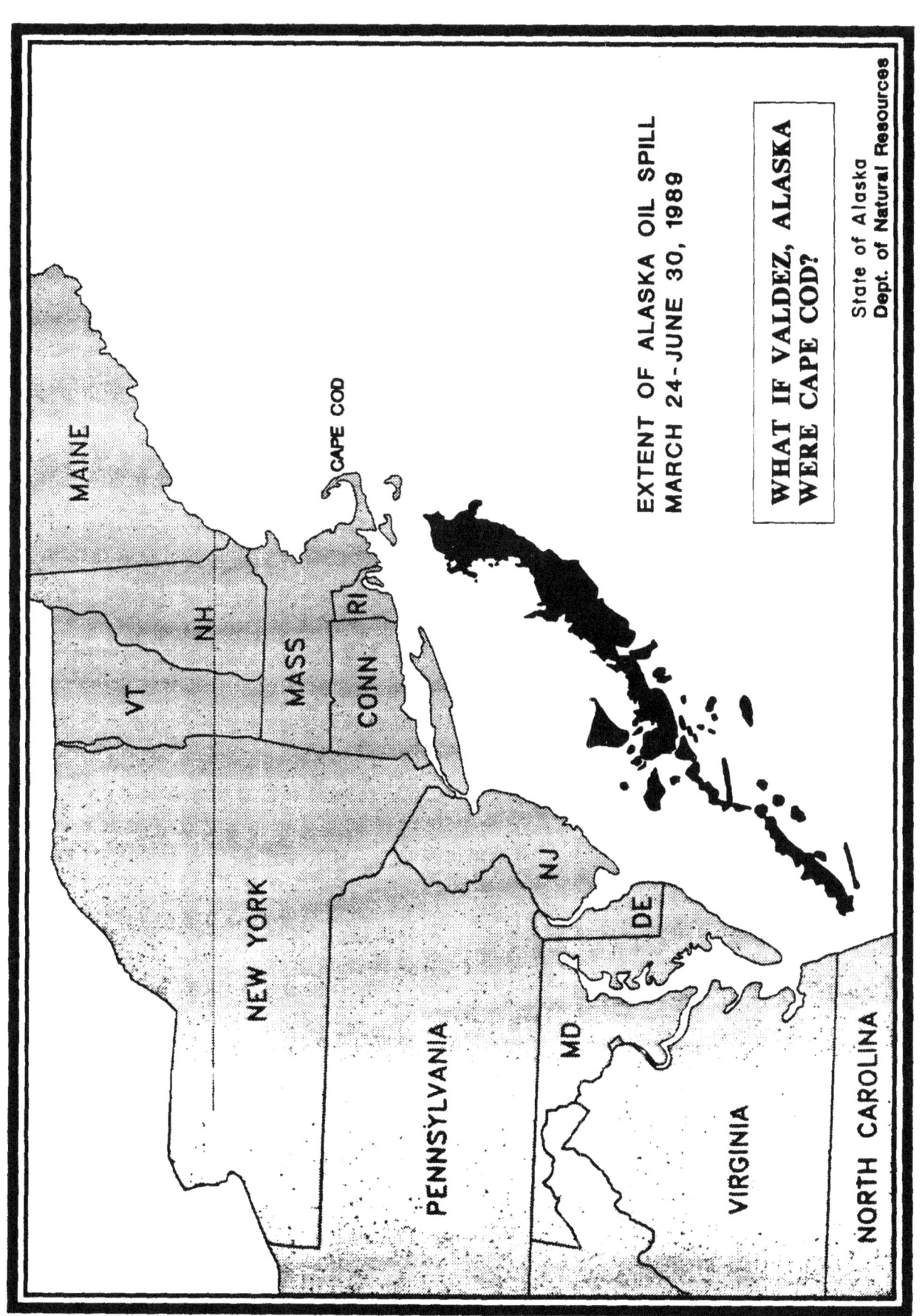

EXTENT OF ALASKA OIL SPILL
MARCH 24-JUNE 30, 1989

WHAT IF VALDEZ, ALASKA
WERE CAPE COD?

State of Alaska
Dept. of Natural Resources

NOTES

1. National Response Team, <u>The Exxon Valdez Oil Spill, A
    Report to the President from Samuel K. Skinner, Secre-
    tary, Department of Transportation and William K.
    Reilly, Administrator, Environmental Protection Agency</u>
    (Washington [?]: May 1989), p. 3. Hereafter cited as
    "National Response Team, <u>The Exxon Valdez Oil Spill, A
    Report to the President</u>."

2. Federal Water Quality Administration, U.S. Department
    of the Interior, "Summary Report: Kodiak Oil
    Pollution, February-March 1970," May 1970, in Oil Spill
    History files, Division of Cultural Resources, Alaska
    Regional Office, National Park Service, hereafter cited
    as Oil Spill History files, NPS (ARO-RCR).

3. Joan M. Antonson and William S. Hanable, <u>Alaska's
    Heritage</u> (Anchorage: The Alaska Historical Society for
    the Alaska Historical Commission, 1985), pp. 432-437;
    <u>The Anchorage Daily News</u>, "Too Little, Too Late,"
    Sunday, October 29, 1989, pp. A-1, A-6 and A-7.

4. Stanley Factor and Sandra J. Grove, "Alaskan transpor-
    tation: an overview of some aspects of transporting
    Alaskan crude oil," in <u>Marine Technology</u> (July 1979) 16
    (3): 211-224; Riki Ott, "Spilled Oil and the Alaska
    Fishing Industry: Looking Beyond Fouled Nets and Lost
    Fishing Time," paper prepared for 1989 Oil Spill
    Conference, San Antonio, Texas, February 14-16, 1989.

5. Alaska Geographic, <u>Where the Mountains Meet the Sea</u>,
    Vol. 13, No. 1 (1986), p. 5.

6. Dr. David Shaw, University of Alaska Institute of
    Marine Sciences, cited in Tim Moffat, "Exxon oil spill
    headed toward Resurrection Bay," <u>The Seward Phoenix
    Log</u>, March 30, 1989, pp. 1, 11; Tom Royers, undated
    note to Senator Stevens on annotated copy of map of
    track of drifting buoy GARS DRIFTER.

7. Hal Bernton, "Oil takes its toll of birds in the Gulf,"
    <u>Anchorage Daily News</u>, April 11, 1989, pp. A1, A8.

8. National Response Team, 1989, p. 27; U.S. Bureau of
    Land Management, <u>Draft Environmental Impact Statement -
    Proposed Outer Continental Shelf Oil and Gas Lease Sale
    - Eastern Gulf of Alaska</u> (Washington [?]: Government
    Printing Office [?], 1979, pp. 137, 139.

9.   Bostwick B. Ketchum, "Oil in the Marine Environment," in National Academy of Sciences, Ocean Affairs Board, <u>Background Papers for a Workshop on Inputs, Fates, and Effects of Petroleum in the Marine Environment</u>, Vol. 2 (Washington, D.C.: 1973), pp. 714-718.

10. Edward Mertens, "A Literature Review of the Biological Impact of Oil Spills in Marine Waters," in National Academy of Sciences, Ocean Affairs Board, <u>Background Papers for a Workshop on Inputs, Fates, and Effects of Petroleum in the Marine Environment</u> (Washington, D.C.: 1973), pp. 745.

11. "Federal Officials Note Persistence of Old Oil," in <u>Oil Spill Chronicle</u>, Vol. 1, No. 14, October 10, 1989, p. 1.

12. Quoted in David Whitney, "Veteran of another spill says cleanup is a long haul," <u>Anchorage Daily News</u>, April 13, 1989, p. A1.

13. Bruce Obee, "Oil Spill Aftermath," <u>Canadian Geographic</u>, April/May 1989, p. 29.

14. National Park Service, "Kenai Fjords National Park," (Washington: Government Printing Office, 1988).

15. Tad Bartimus, "Kenai Fjords, whales, threatened by spill," <u>The Anchorage Times</u>, April 4, 1989.

16. Alaska Planning Group, National Park Service, <u>Final Environmental Impact Statement: Proposed Harding Icefields - Kenai Fjords National Monument, Alaska</u>, 1974, pp. 59-60.

17. David Damas, Ed., <u>Handbook of North American Indians, Vol. 5, Arctic</u> (Washington, D. C.: Smithsonian Institute, 1984), pp. 198-199.

18. National Park Service, "Lake Clark," (Washington, D.C.: Government Printing Office, 1985).

19. Alaska Geographic Society, <u>Lake Clark-Iliamna</u>, <u>Alaska Geographic</u> 13 (4) (1986), p. 93; National Park Service, "Lake Clark," 1985.

20. Damas, pp. 198-199.

21. National Park Service, "Katmai," (Washington, D.C.: Government Printing Office, 1988).

22. Damas, pp. 74-75.

23. David Ames, June 2, 1989, memorandum to Jack Morehead, Subject: Oil Spill Background For You, in "Miscel- laneous Documents" file, Oil Spill History files, NPS (ARO-RCR).

24. Paul F. Haertel, October 18, 1989, interview with William S. Hanable, in Oil Spill History files, NPS (ARO-RCR).

25. Paul D. Gates, and Pamela A. Bergmann, July 5, 1989, interview with William S. Hanable, interview record in Oil Spill History files, NPS (ARO-RCR).

26. National Response Team, <u>A Report on the National Oil and Hazardous Substances Response System - Annual Report 1988</u> (Washington, D.C.: March 6, 1989); National Response Team, <u>The Exxon Valdez Oil Spill, A report to the President</u>, pp. i, 5.

27. Haertel, October 17, 1989. William B. Lawrence, Chief, Environmental Compliance, Alaska Regional Office, National Park Service, October 13, 1989, interview with William S. Hanable, in Oil Spill History files, NPS (ARO-RCR).

28. Lawrence, October 13, 1989.

29. Gates and Bergmann, July 5, 1989.

30. Dave Liebersbach, Seward Incident Command Team Commander, April 14, 1989, interview with William S. Hanable, in Oil Spill History files, NPS (ARO-RCR).

31. Don Cornett, Memorandum, To Whom It May Concern, April 6, 1989, in "Miscellaneous Situation Reports" file, Oil Spill History files, Alaska Regional Office, National Park Service (hereafter referred to as NPS (ARO-RCR).

32. Gates and Bergmann, July 5, 1989.

33. Liebersbach, April 14, 1989.

34. Plans, Seward ICT, "Incident Narrative: Exxon Valdez Oil Spill, Seward Incident Command," April 16, 1989, in Plans Package, Seward ICT files.

35. Homer Branch, Seward Incident Command Team, "Seward Oil Spill - Incident Command Post (ICP - Homer) - Transi- tion Briefing Package, in "Homer Transition Original" file, in Plans Package, Kenai and Homer file, Seward ICT files.

36. Plans, Seward ICT, "Incident Narrative: Exxon Valdez Oil Spill, Seward Incident Command," April 16, 1989.

37. Gates and Bergmann, July 5, 1989.

38. Liebersbach, April 14, 1989.

39. William "Bud" D. Rice, Resource Management Specialist, Kenai Fjords National Park, June 15, 1989, interview with William S. Hanable, in Oil Spill History files, NPS (ARO-RCR).

40. Anne D. Castellina, Superintendent, Kenai Fjords National Park, June 15, 1989, interview with William S. Hanable, in Oil Spill History files, NPS (ARO-RCR).

41. David B. Ames, Associate Regional Director for Operations, Alaska Region, National Park Service, June 9, 1989, interview with William S. Hanable, in Oil Spill History files, NPS (ARO-RCR); Castellina, June 15, 1989.

42. David B. Ames, August 3, 1989, interview with Leland J. Shackleton, in "Case Incident Record: Oil Tanker Spill, Case AK-NPS-01," in files of Alaska Regional Office, National Park Service.

43. Ames, June 9, 1989.

44. Richard G. O'Guin, Chief, Protection and Ranger Activities Division, Alaska Regional Office, National Park Service, April 3, 1989, personal communication to William S. Hanable.

45. Castellina, June 15, 1989.

46. Dennis P. Galvin, Deputy Director, National Park Service, May 22, 1989, interview with William S. Hanable, in Oil Spill History files, NPS (ARO-RCR).

47. Galvin, May 22, 1989.

48. Galvin, May 22, 1989.

49. Don Hamburger, "Criticism in oil spill case now spreading to Interior Department," in "Other Press Releases" file, Oil Spill Information Office file, Plans Package, Seward ICT files.

50. Galvin, May 22, 1989.

51. Boyd Evison, Alaska Regional Director, National Park Service, October 17, 1989, interview with William S. Hanable, in Oil Spill History files, NPS (ARO-RCR).

52. Evison, October 17, 1989.

53. Evison, October 17, 1989.

54. Evison, October 17, 1989.

55. Evison, October 17, 1989.

56. "Criticism in oil spill case now spreading to Interior Department," April 13, 1989, in [Washington] Star Tribune, clipping in "Other Press Releases" file, Seward ICT files.

57. "A big story to tell," The Anchorage Times, April 11, 1989, p. B-4.

58. Sen. Ted Stevens, January 5, 1990, interview with William S. Hanable, in Oil Spill History files, NPS (ARO-RCR).

59. Sen. Ted Stevens, Newsletter, April 1989.

60. Castellina, June 15, 1989; Ranger Activities Division, Washington Office, National Park Service, "Morning Report," March 31, 1989; Cordell Roy, April 1, 1989.

61. Evison, October 17, 1989.

62. Plans, Seward Incident Command Team, "Incident Narrative: Exxon Valdez Oil Spill, Seward Incident Command," April 16, 1989, ms. in Plans Package, Seward ICT files.

63. Anne D. Castellina, Superintendent, Kenai Fjords National Park, March 30, 1989, to Dave Liebersbach, Incident Commander, in files of Kenai Fjords National Park.

64. Castellina, June 15, 1989; Plans, Seward ICT, "Incident Narrative: Exxon Valdez Oil Spill, Seward Incident Command," April 16, 1989.

65. Rice, June 15, 1989.

66. Rice, June 15, 1989.

67. Spencer, August 9, 1989.

68. Rice, June 15, 1989.

69. Bud Rice, 03/03/89, Significant Salmon Streams in Descending Order of Priority for Protection in the Resurrection Bay/Gulf Coast Area Near Seward, in Plans Package, Seward ICT files.

70. Joseph P. Stam, November 7, 1989, interview with William S. Hanable, in Oil Spill History files, NPS (ARO-RCR).

71. Bud Rice, 03/31/89, Revised Significant Salmon Streams in Descending Order of Priority for Protection in the Resurrection Bay/Gulf Coast Area Near Seward, in Plans Package, Seward ICT files.

70. Anne D. Castellina, "Superintendent's (Line Officer's) Briefing Package," March 29, 1989, in "Delegation of Authority" file, Plans Package, Seward ICT files.

71. Castellina, March 29, 1989; Anne D. Castellina, Superintendent, Kenai Fjords National Park, March 30, 1989, "Assignment of Land Management Agency Representative," to Dave Liebersbach, Incident Commander, in "Delegation of Authority" file, Seward ICT files.

72. Castellina, March 29, 1989.

73. Castellina, June 15, 1989.

74. Plans, Seward ICT, "Incident Narrative: Exxon Valdez Oil Spill, Seward Incident Command," April 16, 1989.

75. Minutes, MAC Group, April 3, 1989, in Oil Spill History files, NPS (ARO-RCR).

76. Castellina, June 15, 1989.

77. MAC Group Minutes, April 12, 1989.

78. Don Gilman, Mayor, Kenai Peninsula Borough, interview with William S. Hanable, August 3, 1989, in Oil Spill History files, NPS (ARO-RCR).

79. Gilman, August 3, 1989.

80. Gilman, August 3, 1989.

81. William S. Hanable, Notes, MAC Group meeting, 0900, 04/04/89, in Oil Spill History files, NPS (ARO-RCR).

82. MAC Group Minutes, April 2, April 3, 1989, in Oil Spill History files, NPS (ARO-RCR).

83. MAC Group Minutes, April 7, 1989.

84. MAC Group Minutes, April 3-15, 1989.

85. MAC Group Minutes, April 10, 1989.

86. Plans, Seward ICT, "Incident Narrative: Exxon Valdez Oil Spill," April 16, 1989.

87. Douglas D. Erskine, August 11, 1989, interview with William S. Hanable, in Oil Spill History files, NPS (ARO-RCR).

88. MAC Group Minutes, April 7, 1989.

89. MAC Group Minutes, April 8, 1989.

90. Plans, Seward ICT, "Incident Narrative: Exxon Valdez Oil Spill, Seward Incident Command," April 16, 1989.

91. MAC Group Minutes, April 11, 1989.

92. Loren Flagg, Chair, Homer MAC Advisory Committee, April 7, 1989, to Seward MAC Team, Table 1, attached to MAC Group Minutes, April 8, 1989.

93. MAC Group Minutes, April 8, 1989.

94. MAC Group Minutes, April 12, April 14, 1989.

95. MAC Group Minutes, April 15, 1989.

96. MAC Group Minutes, April 13, 1989.

97. Regional Response Team, "RRT Incident 34: T/V Exxon Valdez, Major Oil Spill, Report 46: OSC Western AK POLREP 1," message with date-time group (DTG) 040700Z Apr 89, in "RRT Briefing Notes" file, Plans Package, Seward ICT files.

98. Capt. Rene Rousell, U.S. Coast Guard, June 19, 1989, interview with William S. Hanable, in Oil Spill History files, NPS (ARO-RCR).

99. Rousell, June 19, 1989.

100. Rousell, June 19, 1989.

101. Erskine, April 11, 1989, personnal communication with William S. Hanable.

102. Rousell, June 19, 1989.

103. Notes from 04/04/89, Alaska Regional Response Team Briefing, in Incident Command Team (PIO) files.

104. William S. Hanable, Notes from 1800 Briefing in Seward, April 6, 1989, in Oil Spill History files, NPS (ARO-RCR).

105. William S. Hanable, Notes from Core Team Meeting, 2030, April 6, 1989, in Oil Spill History files, NPS (ARO-RCR).

106. O'Guinn, 0900, April 3, 1989.

107. Quoted in Castellina, June 15, 1989.

108. Anne D. Castellina, Superintendent, Kenai Fjords National Park, "Kenai Fjords and the Exxon Valdez Oil Spill - An Initial Report," April 29, 1989, in Oil Spill History files, NPS (ARO-RCR).

109. Castellina, June 15, 1989.

110. Shift Plan for 04/03/89, prepared 04/02/89, in Plans Package, Seward ICT files.

111. Janis M. Meldrum, September 14, 1989, interview with William S. Hanable, in Oil Spill History files, NPS (ARO-RCR).

112. Roy, November 14, 1989.

113. G. Ray Bane, April 18, 1990; Daniel M. Hamson, December 8, 1989.

114. G. Ray Bane, September 14, 1989.

115. Cordell Roy, "Resource Risk Assessment - Katmai National Park and Preserve," April 4, 1989, in Environmental Reports file, Plans Package, Seward ICT files.

116. Janis M. Meldrum, Resource Management Specialist, Katmai National Park and Preserve/Aniakchak National Monument and Preserve, May 15, 1990, memorandum to Acting Chief, Division of Cultural Resources, Anchorage Regional Office, Subject:

Review of "The <u>Exxon Valdez</u> Oil Spill: The National Park
Service Response."

117. Meldrum, May 15, 1990.

118. Bane, September 14, 1989.

119. Bane, September 14, 1989.

120. A. E. Hutchison, Superintendent, Lake Clark National Park
and Preserve, April 5, 1989, memorandum to Incident
Commander, Seward, Subject: Action Plan, in "Memos" file,
Plans Package, Seward ICT files.

121. MAC Group, Minutes, April 5, 1989.

122. Stephen M. Hurd, Acting Superintendent, Katmai National
Park and Preserve/Aniakchak National Monument, April 6,
1989, Memorandum to Associate Regional Director, Opera-
tions, Alaska Regional Office, National Park Service,
Subject: Request to Activate Incident Command Team, in
"Delegation of Authority" file, Plans Package, Seward
ICT files.

123. Bane, September 14, 1989.

124. Peter Fitzmaurice, Acting Chair, MAC Group, Seward,
April 6, 1989, Memorandum to Incident Management Team,
Plans/Logistics, Subject: Actions Items as a Result of
4/6/89 Meeting, in Oil Spill History files.

125. Plans, Seward ICT, "Incident Narrative: Exxon Valdez
Oil Spill, Seward Incident Command," April 16, 1989.

126. Bane, April 18, 1990.

127. William S. Hanable, Notes from Core Team Meeting, 2030,
April 6, 1989, in Oil Spill History files, NPS (ARO-
RCR).

128. William S. Hanable, Notes from Plans Section meeting,
1130, April 6, 1989, in Oil Spill History files (ARO-
RCR).

129. William S. Hanable, Notes from 1800 Briefing, Seward
ICT, April 7, 1989, in Oil Spill History files; Homer
Branch, ICT, "Incident Narrative," April 15, 1989, in
Plans Package, Seward ICT files; Plans, Seward ICT,
"Incident Narrative: Exxon Valdez Oil Spill," April
16, 1989.

130. Stephen M. Hurd, Chief Ranger, Katmai National Park and Preserve, interview, September 14, 1989, with William S. Hanable, in Oil Spill History files, NPS (ARO-RCR).

131. Homer Branch ICT, Incident Narrative, April 15, 1989.

132. Plans, Seward ICT, "Incident Narrative: Exxon Valdez Oil Spill, Seward Incident Command," April 16, 1989.

133. Dave Liebersbach, April 17, 1989, interview with Frank Dean, in "Incident Case Record: Oil Tanker Spill, AK-NPS-01."

134. Plans, Seward ICT, "Incident Narrative: Exxon Valdez Oil Spill, Seward Incident Command," April 16, 1989.

135. Kenai Branch ICT, "Incident Narrative," undated, in Plans Package, Seward ICT files.

136. Plans, Seward ICT, "Incident Narrative: Exxon Valdez Oil Spill, Seward Incident Command," April 16, 1989.

137. MAC Group Minutes, April 11, 1989.

138. Erskine, August 11, 1989.

139. Plans, Seward ICT, "Incident Status Summary," March 31, 1989.

140. Shift Plan for 04/01/89, prepared 03/31/89, in Plans Package, Seward files; William S. Hanable, Notes from 04/04/0830/89 Plans Section meeting, in Oil Spill History files, NPS (ARO-RCR).

141. Bureau of Land Management Fax Transmission, April 2, 1989, in Plans Package, Seward ICT files.

142. Shift Plan for 04/02/89, prepared 04/01/89, in Plans Package, Seward files.

143. Planning Section Chief, 04/02/89, to Section Chiefs, Seward Incident, in Plans Package, Seward ICT files.

144. William S. Hanable, Notes from briefing by David B. Ames Acting Director, Alaska Region, at Alaska Regional Office, National Park Service, 0900, April 3, 1989, in Oil Spill History files, NPS (ARO-RCR).

145. William S. Hanable, Notes from briefing by Richard O. O'Guinn, Chief, Ranger Protection Activities, Alaska Region, National Park Service, at 0900 briefing, April 3, 1989, in Oil Spill History files, NPS (ARO-RCR).

146. Cordell Roy, April 1, 1989, Notes from telephone conference, 1415, April 1, 1989, in "ARRT Briefing Notes"
files, Situations Unit file, Plans Package, Seward ICT
files.

147. Paul Gleeson, October 16, 1989, interview with William
S. Hanable, in Oil Spill History files, NPS (ARO-RCR).

148. Gleeson, October 16, 1989.

149. O'Guinn, April 3, 1989, personal communication to
William S. Hanable.

150. Art Latterell, "Report on Basic ICS Training," April
10, 1989, memo to Doug Erskine, NPS, Alaska Regional
Fire Management Officer, in "ICT Trainees" file, Plans
Package, Seward ICT files.

151. Incident Shift Plan, 04/04/89, prepared 04/03/89, in
Plans Package, Seward ICT files.

152. Plans, Seward Incident Command Team, "Incident Narrative: Exxon Valdez Oil Spill, Seward Incident
Command," April 16, 1989.

153. Plans, Seward Incident Command Team, "Incident
Narrative: Exxon Valdez Oil Spill, Seward Incident
Command," April 16, 1989.

154. Plans, Seward ICT, "Incident Narrative: Seward
Incident Command," April 16, 1989.

155. Don Fuller, Logistics Section Chief, April 11, 1989, to
Anne Castellina, MAC Chairwoman, attached to MAC Group
Minutes, April 11, 1989.

156. Ronald Knowles, April 13, 1989, interview with Scott
Taylor, in "Case Incident Record: Oil Tanker Spill,
AK-NPS-01," in files of Alaska Regional Office,
National Park Service.

157. Operations, Seward ICT, Planning Worksheet, March 31,
1989, in Operations Package, Seward ICT files.

158. "Marine Mammal Survey Notes," April 6, 1989, in "Pre-
Spill Impact" file, Situation Unit file, Plans Package,
Seward ICT files; Operations, Seward ICT, March 31,
1989.

159. Operations, Seward ICT, March 31, 1989.

160. Operations, Seward ICT, March 31, 1989; Beach Survey Crew [Team "D"], "Log of Observations of Beach Survey Crew--Foxy Lady, Belinda Bain & Don Dragoo, in "Field Beach Survey Pre-Spill Impact" file in Situations Unit files, Plans Package, Seward ICT files.

161. Operations, Seward ICT, Planning Worksheet, April 2, 1989, in Operations Package, Seward ICT files.

162. The Anchorage Times, "Fire experts lend aid in keeping oil from harming park," April 3, 1989, press clipping in "Other Press Releases," file, Seward ICT files.

163. William D. Rice, "Data Collection Plan," April 3, 1989, in Seward ICT files; Operations, Seward ICT, Planning Worksheet, April 4, 1989, in Operations Package, Seward ICT files; Plans, Seward ICT, "Incident Status Summary," April 5, 1989, in Plans Package, Seward ICT files.

164. Operations, Seward ICT, Planning Worksheets, April 8, 1989, April 10, 1989, in Operations Package, Seward ICT files; Plans, Seward ICT, "Incident Status Summary," April 9, 1989, in Plans Package, Seward ICT files.

165. Plans, Seward ICT, "Incident Status Summary," April 11, 1989, in Plans Package, Seward ICT files.

166. Acting Superintendent, Katmai National Park and Preserve/Aniakchak National Monument, April 6, 1989, Memorandum to Associate Director, Operations, Alaska Regional Office, National Park Service, Subject: Request to Activate ICT, in "Delegation of Authority" file, Plans Package, Seward ICT.

167. A. E. Hutchison, Superintendent, Lake Clark National Park and Preserve, April 5, 1989, Memorandum to Incident Commander, Spill Seward, Subject: Action Plan, in "Memos" file, Plans Package, Seward ICT.

168. Operations Section, Kenai Branch ICT, "Planning Worksheet," April 9, 1989, in Plans Package - Kenai and Homer, Seward ICT files.

169. Kenai Branch, Seward ICT, "Incident Narrative," undated, in Plans Package, Seward ICT files; Patricia L. McClenahan, interview with Miller, in Incident Case Record, "Oil Tanker Spill," AK-NPS-01.

170. Homer Branch ICT, "Incident Narrative," April 15, 1989, in Plans Package, Seward ICT files.

171. Garey Coatney, June 1989, interview with William S. Hanable, in Oil Spill History files, NPS (ARO-RCR).

172. Plans, "Incident Status Summary," [for dates indicated], in Plans Package, Seward ICT files.

173. Coatney, June 1989.

174. Rae Baxter, "Katmai National Park - Survey Report of 15-29 April 1989," in files of Katmai National Park and Preserve.

175. William S. Hanable, Notes on briefing by Anne D. Castellina, 1900, April 8, 1989, remarks to public meeting at Seward Vocational-Technical School, in Oil Spill History files, NPS (ARO-RCR).

176. Plans, Seward ICT, "Incident Status Summary," April 2, 1989, in Plans Package, Seward ICT files.

177. Plans, Seward ICT, "Incident Narrative: Exxon Valdez Oil Spill, Seward Incident Command," April 16, 1989.

178. Minutes, MAC Group, April 3, 1989, in Oil Spill History files, NPS (ARO-RCR).

179. William S. Hanable, Notes from Incident Command Team Briefing, 04/03/1800, in Oil Spill History files, NPS (ARO-RCR).

180. Dave Liebersbach, April 3, 1989.

181. William S. Hanable, Notes on briefing by John Gage, City of Seward Fire Chief and Co-Commander of Incident Command Team, April 3, 1989, in Oil Spill History files.

182. Operations, Seward ICT, Planning Worksheet, April 3, 1989, in Operations Package, Seward ICT files.

183. Capt. Rene Rousell, U.S. Coast Guard, June 19, 1989, interview with William S. Hanable, in Oil Spill History files, NPS (ARO-RCR).

184. U.S. Coast Guard Marine Safety Office, Anchorage, Message, 050435Z April 1989, Subject: POLREP 2, Major Oil Spill, T/V Exxon Valdez, Bligh Reef, Alaska, in Oil Spill History files, NPS (ARO-RCR), hereafter referred to as POLREP 2.

185. Minutes, MAC Group, April 4, 1989, in Oil Spill History files, NPS (ARO-RCR).

186. Minutes, MAC Group Meeting, April 4, 1989.

187. Plans, Seward ICT, "Incident Narrative: Exxon Valdez Oil Spill, Seward Incident Command," April 16, 1989.

188. Dr. Ron Goodman, Minutes, MAC Group, April 5, 1989, in Oil Spill History files, NPS (ARO-RCR).

189. Minutes, MAC Group, April 5, 1989.

190. Minutes, MAC Group, April 6, 1989, in Oil Spill History files, NPS (ARO-RCR).

191. Minutes, MAC Group, April 7, 1989, in Oil Spill History files, NPS (ARO-RCR).

192. Plans, Seward ICT, "Incident Narrative: Exxon Valdez Oil Spill, Seward Incident Command," April 16, 1989.

193. In "Environmental Reports" file, Plans Package, Seward ICT files.

194. Homer Branch, ICT, "Homer ICP Narrative," April 16, 1989, in Plans files, Seward ICT; Plans, Seward ICT, "Incident Status Summary," April 10, 1989.

195. Plans, Seward ICT, "Incident Status Summary," April 11, 1989, in Plans Package, Seward ICT files.

196. Operations, Planning Worksheet, April 12, 1989, in Operations Package, Seward ICT files.

197. Minutes, MAC Group, April 12, 1989, in Oil Spill History files, NPS (ARO-RCR).

198. Minutes, MAC Group, April 12, 1989, in Oil Spill History files, NPS (ARO-RCR); Plans, Seward ICT, "Incident Narrative," April 13, 1989.

199. Minutes, MAC Group, April 13, 1989, in Oil Spill History files, NPS (ARO-RCR).

200. Operations, Seward ICT, "Oil Spill - Seward," April 11, 1989, in Operations Package, Seward ICT files; Plans, Incident Status Summary," April 12, 1989, in Plans Package, Seward ICT files.

201. Joseph C. Stam, Branch Director, Homer ICT Operations, November 6, 1989, interview with William S. Hanable, in Oil Spill History files, NPS (ARO-RCR).

202. Evison, October 17, 1989.

203. Regional Director, Alaska Regional Office, National Park Service, May 8, 1989, Memorandum to Director, National Park Service, Subject: Oil Spill Assistance for Alaska Region, in Alaska Area Command ICT files.

204. Boyd Evison, "Limited Delegation of Authority," to John Kraushaar, May 11, 1989, in Alaska Area Command ICT files.

205. Evison, May 11, 1989.

206. Evison, "Line Manager's Briefing," May 11, 1989, in Alaska Area Command ICT files.

207. Evison, October 17, 1989.

208. James A. Randall, September 27, 1989, interview with William S. Hanable, in Oil Spill History files, NPS (ARO-RCR).

209. Frank J. Betts, September 29, 1989, interview with William S. Hanable, in Oil Spill History files, NPS (ARO-RCR).

210. Castellina, June 15, 1989.

211. Erskine, August 11, 1989.

212. Seward ICT, Incident Action Plan, April 16, 1989, in Plans Package, Seward ICT files.

213. Castellina, June 15, 1989.

214. Castellina, June 15, 1989.

215. Kenai Fjords ICT, Incident Action Plan, May 1, 1989, in Alaska Area Command ICT files.

216. Page Spencer, Trip Report M/V Spirit, Confirmation of Oil Exposure, Kenai Fjords National Park, May 5, 1989, in Alaska Area Command ICT files.

217. Spencer, May 5, 1989.

218. Castellina, June 15, 1989.

219. Anne D. Castellina, Superintendent, Kenai Fjords National Park, November 3, 1989, interview with William S. Hanable, in Oil Spill History files, NPS (ARO-RCR).

220. Dwayne D. Atwood, "Navy skimmers here to pick up Exxon's spilled oil," in The Seward Phoenix Log, April 13, 1989, p. 1; Seward ICT, News Update, April 11, 1989, in Public Information folder, Plans Package, Seward ICT files; U.S. Coast Guard, "Exxon Valdez Fact Sheet," April 11, 1989, in Plans Package, Seward ICT files.

221. Fred Bayles, "Weather breaks up oil slick," in The Anchorage Times, April 12, 1989, pp. A1, A10.

222. Seward MAC Group, Minutes, April 11, 1989, in Plans Package, Seward ICT files.

223. U.S. Coast Guard, April 11, 1989; Plans, Seward ICT, Incident Narrative, April 12, 1989.

224. U.S. Coast Guard, Message, no date-time group, POLREP 11, Subject: Major Oil Spill, T/V Exxon Valdez, Blight Reef, AK, in "Coast Guard Updates" file, Plans Package, Seward ICT files; Regional Response Team, "Briefing Notes," April 13, 1989, in "RRT Briefing Notes" file, Plans Package, Seward ICT files.

225. Plans, Seward ICT, Incident Narrative, April 16, 1989.

226. Tim Moffatt, "Exxon oil spill headed toward Resurrection Bay," in The Seward Phoenix Log, April 13, 1989, p. 1.

227. U.S. Coast Guard cutter Yocona, Message, 152142Z April 1989, to Coast Guard Marine Safety Office, Anchorage, Subject: Marine Environmental Response SITREP 10: T/V Exxon Valdez Aground, and Message 162020Z April 1989, SITREP 11, same subject, same addressee, in "Coast Guard Updates" file, Plans Package, Seward ICT files.

228. Castellina, June 15, 1989.

229. Castellina, November 3, 1989.

230. Castellina, November 3, 1989.

231. Castellina, November 3, 1989; August 25, 1989, letter to D.D. Emmal, President, The English Bay Corporation, in Alaska Area Command ICT files.

232. Castellina, November 3, 1989.

233. Castellina, November 3, 1989.

234. Kenai Fjords ICT, Incident Action Plan, May 16, 1989, in files of the Alaska Area Command ICT.

235. Kenai Fjords ICT, Incident Action Plan, April 24, 1989.

236. Castellina, November 6, 1989.

237. Kenai Fjords ICT, Incident Action Plan, June 1, 1989, in Alaska Area Command ICT files.

238. Garey Coatney, June 5, 1989, Memorandum to Frank Betts, Area Command, ARO, Subject: Projected Operational Activities and Associated Costs Oil Spill Response Management Plan, Kenai Fjords NP, in Alaska Area Command ICT files.

239. Kenai Fjords ICT, Incident Action Plan, July 1, 1989, in Alaska Area Command ICT files.

240. Russell Kucinski, July 12, 1989, Memorandum to Frank Betts, Garey Coatney, Anne Castellina, Subject: Meeting with Exxon, 7-12-89, in Alaska Area Command ICT files.

241. Kenai Fjords ICT, Incident Status Summary, August 1, 1989, in Alaska Area Command ICT files.

242. Anne D. Castellina, Superintendent, Kenai Fjords National Park, August 17, 1989, Superintendent's (Line Officer's) Briefing, Exxon Valdez Oil Spill Response, Phase II and Phase III, in Alaska Area Command ICT files.

243. G. Ray Bane, April 18, 1990, memorandum to Acting Chief, Division of Cultural Resources, Alaska Region, Subject: "The Exxon Valdez Oil Spill: The National Park Service Response" - Review.

244. Homer ICT, Incident Narrative, April 14, 1989; Bane, September 14, 1989.

245. Katmai National Park and Preserve, Long-term Monitoring of Oiled Beaches, Katmai National Park and Aniakchak National Preserve, June 18, 1989, briefing in files of Katmai National Park and Preserve.

246. Nancy Deschu, April 26, 1989, field notebook.

247. Cordell Roy, Environmental Specialist, Alaska Regional Office, National Park Service, November 14, 1989, interview with William S. Hanable, in Oil Spill History files, NPS (ARO-RCR).

248. Kodiak ICT, Incident Action Plan, April 16, 1989; Incident Status Summary, April 17, 1989; Incident Status Summary, April 18, 1989, in Alaska Area Command ICT files.

249. Kodiak ICT, Incident Status Summary, April 22, 1989; Roy, November 14, 1989.

250. Kodiak ICT, Incident Status Summary, May 2, 1989, in Alaska Area Command files; Steve Rinehart, "Oil takes its toll at Katmai," Anchorage Daily News, May 3, 1989, pp. A1, A10.

251. Bane, September 14, 1989; Kodiak ICT, Incident Status Summary, May 1, 1989, in Alaska Area Command ICT files; Roy, November 14, 1989.

252. Kodiak ICT, Vessel/Aircraft Log, May 2, 1989, in Alaska Area Command ICT files.

253. Roy, November 14, 1989.

254. Bane, September 14, 1989.

255. Bane, September 14, 1989; Seward MAC Group, Minutes, May 8, 1989.

256. Gilbert B. Blinn, Superintendent, Lassen Volcanic National Park, November 4, 1989, interview with William S. Hanable, in Oil Spill History files, NPS (ARO-RCR).

257. Bane, September 14, 1989.

258. Kodiak ICT, Incident Status Summary, May 19, 1989, in Alaska Area Command ICT files; Superintendent, Katmai National Park and Aniakchak National Monument, May 28, 1989, Memorandum to Regional Director, Alaska Region, National Park Service, et al., Subject: Oil Spill Clean-up Evaluation, in Alaska Area Command ICT files.

259. Kodiak ICT (name changed to Katmai Field Office at the end of May to avoid confusion with the Coast Guard/Exxon Incident Command Post at Kodiak), Vessel Log, June 24 and 25, 1989, in Alaska Area Command ICT files.

260. Chief Biologist [Will Troyer], Kodiak Field Office, July 27, 1989, Memorandum to Incident Commander, Kodiak Field Office, Subject: Priorities for Shoreline Cleanup EXXON VALDEZ Oil Spill, in Kodiak ICT files.

261. Katmai Field Office, Vessel Log, August 6, 1989, August 15, 1989, September 15, 1989; Will Troyer, 1990 Impact Studies and Monitoring Plan, Exxon Valdez Oil Spill, Katmai National Park, September 11, 1989, in Kodiak ICT files, Alaska Area Command ICT files.

262. Janis M. Meldrum, Resource Management Specialist, Katmai National Park and Preserve, September 14, 1989, interview with William S. Hanable, in Oil Spill History files, NPS (ARO-RCR).

263. Kodiak Field Office, Katmai National Park and Preserve, "1989 Operations and Monitoring Plan Exxon Valdez Oil Spill," in files of Katmai National Park and Preserve.

264. Janis M. Meldrum, May 15, 1990.

265. Meldrum, September 14, 1989.

266. Kodiak ICT, Incident Status Summary, June 19, 1989, in Kodiak ICT files.

267. Incident Commander, Kodiak Incident Command, June 11, 1989, Memorandum to Logistics Chief, Anchorage Area Command, Subject: Selection of Charter Vessel for Resource Monitoring Program, in Kodiak ICT files.

268. Daniel M. Hamson, Environmental Specialist, Alaska Regional Office, National Park Service, December 8, 1989, interview with William S. Hanable, in Oil Spill History files, NPS (ARO-RCR).

269. Chief Biologist, Kodiak, July 30, 1989, Memorandum to Incident Commander, Betts, Anchorage, Subject: Permanent Monitoring Plots, Katmai and Aniakchak, in Kodiak ICT files.

270. Will Troyer, July 31-August 4, 1989, in Kodiak ICT files.

271. Will Troyer, "1989 Impact Studies and Monitoring Plan Exxon Valdez Oil Spill, Katmai National Park," August 12, 1989, in Kodiak ICT files.

272. Troyer, August 12,1989.

273. Janis M. Meldrum, May 15, 1990.

274. Will Troyer, "1990 Impact Studies and Monitoring Plan, Exxon Valdez Oil Spill, Katmai National Park," September 11, 1989, in Kodiak ICT files.

SELECTED BIBLIOGRAPHY

Alaska Area Command ICT files, Alaska Regional Office, National Park Service.

Alaska Geographic Society, Lake Clark-Iliamna, Alaska Geographic 13 (4) (1986).

Ames, David B., June 9, 1989, interview with William S. Hanable.

Atwood, Dwayne D., "Navy skimmers here to pick up Exxon's spilled oil," in The Seward Phoenix Log, April 13, 1989.

Bane, G. Ray, September 14, 1989, interview with William S. Hanable.

Bayles, Fred, "Weather breaks up oil slick," in The Anchorage Times, April 12, 1989, p. A1, A10.

Bernton, Hal, "Oil takes its toll of birds in the Gulf," Anchorage Daily News, April 11, 1989, p. A1, A8.

Betts, Frank J., September 29, 1989, interview with William S. Hanable.

Blinn, Gilbert B., November 4, 1989, interview with William S. Hanable.

Castellina, Anne D., June 15 and November 3, 1989, interviews with William S. Hanable.

Coatney, Garey, June 1989, interview with William S. Hanable.

Deschu, Nancy, April 1989, field notebook.

Erskine, Douglas D., August 11, 1989, interview with William S. Hanable.

Evison, Boyd, October 17, 1989, interview with William S. Hanable.

Factor, Stanley and Sandra J. Grove, "Alaskan transportation: an overview of some aspects of transporting Alaskan crude oil," in Marine Technology (July 1979) 16 (3): 211-224.

Federal Water Quality Administration, U.S. Department of the Interior, "Summary Report: Kodiak Oil Pollution, February-March 1970," May 1970.

Galvin, Dennis P., May 22, 1989, interview with William S. Hanable.

Gates, Paul D. and Pamela A. Bergmann, July 5, 1989, interview with William S. Hanable.

Gilman, Don, August 3, 1989, interview with William S. Hanable.

Gleeson, Paul, October 16, 1989, interview with William S. Hanable.

Hurd, Stephen M., September 14, 1989, interview with William S. Hanable.

Ketchum, Bostwick, "Oil in the Marine Environment," in National Academy of Sciences, Ocean Affairs Board, Background Papers for a Workshop on Inputs, Fates, and Effects of Petroleum in the Marine Environment (Washington: 1973), pp. 709-725.

Lawrence, William, October 13, 1989, interview with William S. Hanable.

Liebersbach, Dave, April 14, 1989, interview with William S. Hanable.

Meldrum, Janis M., September 14, 1989, interview with William S. Hanable.

Moffatt, Tim, "Exxon oil spill heads toward Resurrection Bay," in The Seward Phoenix Log, April 13, 1989, p. 1.

National Park Service

　　"Case Incident Record: Oil Tanker Spill, Case AK-NPS-01," in Alaska Regional Office, National Park Service.

　　"Katmai," (Washington: Government Printing Office, 1988). Leaflet.

　　"Kenai Fjords National Park," (Washington: Government Printing Office, 1988). Leaflet.

　　"Lake Clark," (Washington: Government Printing Office, 1985). Leaflet.

National Response Team

　　The Exxon Valdez Oil Spill, A Report to the President from Samuel K. Skinner, Secretary, Department of Transportation, and William K. Reilly, Administrator, Environmental Protection Agency (Washington: May 1989).

A Report on the National Oil and Hazardous Substances
Response System - Annual Report 1988 (Washington:
March 6, 1989).

Obee, Bruce, "Oil Spill Aftermath," Canadian Geographic,
April/May 1989, 109 (102).

Oil Spill Chronicle.

Oil Spill History files, Alaska Regional Office, National Park
Service.

Ott, Riki, "Spilled Oil and the Alaska Fishing Industry: Looking
Beyond Fouled Nets and Lost Fishing Time," paper prepared for
1989 Oil Spill Conference, San Antonio, Texas, February 14-16,
1989.

Plans Package, Homer and Kenai Incident Command Teams, Kenai
Fjords National Park.

Plans Package, Seward Incident Command Team, Kenai Fjords
National Park.

Randall, James A., September 27, 1989, interview with William S.
Hanable.

Rice, William "Bud" D., June 15, 1989, interview with William S.
Hanable.

Rousell, Capt. Rene, U.S. Coast Guard, June 19, 1989, interview
with William S. Hanable.

Roy, Cordell, November 14, 1989, interview with William S.
Hanable.

Spencer, Page, August 9, 1989, interview with William S. Hanable.

Stam, Joseph C., November 6, 1989, interview with William S.
Hanable.

U.S. Bureau of Land Management, Draft Environmental Impact
Statement - Proposed Outer Continental Shelf Oil and Gas Lease
Sale - Eastern Gulf of Alaska (Washington: Government Printing
Office, 1979.

Whitney, David, "Veteran of another spill says cleanup is a long
haul," Anchorage Daily News, April 13, 1989, p. A1.

Appendix

KEY PERSONNEL LIST
EXXON VALDEZ OIL SPILL RESPONSE

Employee Name	Position	Location	Responsibility	Park	Agency
ADKISSON	KEN	CH RANGER, BELA	KEFJ	RANGER	BELA	NPS
AHLSTRAND	GARY	INTERTIDAL/VEGETATION	ANCH	INTERTIDAL/VEGETATION	ARO	NPS
ALBERT	DAVID		LACL	BIO TECH	DENA	NPS
ALDERSON	JUDY	RESOURCE MANAGER	KATM	RESOURCE MANAGEMENT	GAAR	NPS
AMENT	KAREN		KEFJ	RPO	GLAC	NPS
AMES	DAVID	ARD,O	ANCH	MANAGEMENT	ARO	NPS
AMUNDSON	GEORGIA	CONTRACTING SPECIALSIT	ANCH	CONTRACTING SPECIALIST	ARO	NPS
ANDERSON	SUSAN	SUPPORT SERV. SUPERVISOR	ANCH	SUPPLIES	ARO	NPS
ANDERSON	PAUL		KEFJ	PLANNING SECTION CHIEF	SHEN	NPS
ANDREWS	CYRIL		KEFJ	ZODIAK OPERATOR		AD
ANDREWS	JANET	CLERK TYPIST	ANCH	CLERK TYPIST	ARO	NPS
ARMOUR	CONLEY		KEFJ	RPO	MACA	NPS
ARMSTRONG	ROBERT		KATM	INCIDENT COMMANDER	NISI	NPS
ARMSTRONG	LISA		KEFJ	DISPATCH		BLM/AFS
ASHLEY	BARBARA	TRAVEL	ANCH	TRAVEL	NOCA	NPS
ASPREY	BRUCE		ANCH	ASST. COMMO. TECH		BLM/AFS
AUSTERMAN	DAWN		KATM	ADMIN CLERK-FINANCE		AD
AXTELL	CRAIG		KATM	RES MGMT SPEC	ROMO	NPS
BABB	BRUCE	SUPPORT DISPATCHER	ANCH	DISPATCH		USFS
BACKES	SALLY	CLERK TYPIST	KATM	CLERK TYPIST	KATM	NPS
BAHE	RALPH	SUPPLY UNIT LEADER	KATM	SUPPLY UNIT LEADER		USFS
BAIN	BELINDA		KATM	BIOLOGIST		AD
BAKER	GERARD		KEFJ	RPO	THRO	NPS
BAKER	CATHERINE		KEFJ	BIOLOGIST	KEFJ	NPS
BANE	RAY	SUPERINTENDENT, KATM	KATM	SUPERINTENDENT	KATM	NPS
BANKS	STEVE	AIR SERV OFF	KATM	AIR SERV OFF		BLM/FSC
BARCUS	BONNY	SUPPORT DISPATCHER	KATM	DISPATCH		BLM/AFS
BARNETT	JIM	DISPATCH	KEFJ	DISPATCH		BLM/AFS
BARNETT	STEVEN		ANCH	WAREHOUSE FOREMAN		AD
BARRETT	MIKE		KEFJ	RPO	NCR	NPS
BAUER	CRAIG	METEOROLOGIST	KEFJ	METEOROLOGIST		BLM/AFS
BAXTER	RAE	BIOLOGIST (MARINE)	LACL	BIOLOGIST (MARINE)	ARO	NPS
BEATTIE	JOAN	RES MGMT	KATM	RES MGMT	DSC	NPS
BEEBE	SUSAN		KEFJ	CLERK TYPIST		AD
BELTON	VERONICA		LACL	FINANCE SECTION CHIEF		BLM/AFS
BENJAMIN	JOHN		KATM	OPERATIONS SECTION CHIEF	GLCA	NPS
BENSON	POPPY	BEACH SURV SPECIALST	KEFJ	BEACH SURVEY		USFWS

Employee Name	Position	Location	Responsibility	Park	Agency
BERENS JIM	ASSOC. REG. DIR., ADMIN.	ANCH	ADMINISTRATION	ARO	NPS
BERNTHAL CHRIS	CONTRACTING SPECIALIST	ANCH	CONTRACTING SPECIALIST	INDU	NPS
BERSON TOM		ANCH	R&D		AD
BERTEMI TERESA	PROCUREMENT SPECIALIST	KEFJ	ICP		BLM
BESSKEN BRUCE		KEFJ	RPO	BADL	NPS
BETTS FRANK	Retired NPS	ANCH	AREA COMMANDER	RETIRED	NPS
BILLER ALLEN	HELICOPTER MGR	KATM	HELICOPTER MANAGER	KATM	BLM/AFS
BIRD FRANK	FISHERIES BIOLOGIST	KATM	BIOLOGIST	KATM	BLM
BIRKEDAL TED	ARCHEOLOGIST	KEFJ	PICHEOLOGIST	ARO	NPS
BLACK JOHN	FILM CREW	KEFJ	FILM CREW	KEFJ	BIFC
BLAIN ROGER		KATM	RPO	ACAD	NPS
BLANK TIM	RANGER	KEFJ	TORT INVESTIGATOR	CURE	NPS
BLASZAK MARSHA	PAYMENT TEAM	KEFJ	PAYMENT TEAM	LAVO	NPS
BLINN GIL		KATM	SUPERINTENDENT'S REP.	LAVO	NPS
BOHAMAN WILLIAM		KEFJ	RPO	FOVA	NPS
BONE STEVEN		KATM	RPO	WICA	NPS
BONGEN ELIZABETH		KATM	MARINE DISPATCHER		AD
BORNEMAN CAROL		KATM	RPO	NCR	NPS
BORTON GORDON		KEFJ	LABORER		AD
BOWKER RANDALL		ANCH	RANGER ACTIVITIES		AD
BOYD JIM	VIDEO CAMERA	KATM	VIDEO EDITOR	GRCA	NPS
BRAGGS JIM		KEFJ	RPO	CANY	NPS
BREEN BOB	PROPERTY OFFICER	KEFJ	PLANNING SEC. CHIEF	ACAD	NPS
BROADWAY DOUG		ANCH	PROPERTY	ARO	NPS
BROADWAY MICHAEL		ANCH	LOGISTICS		AD
BROCK MAC		KEFJ	LAB TECH/BIOLOGIST	GRBA	NPS
BROMSON JERRY		KATM	BIO TECH	KATM	NPS
BROWN EVA	FINANCE SPECIALIST	KEFJ	FINANCE		BLM-AFS
BROWNLEE JEFF	BIOLOGIST	KEFJ	BIOLOGIST		AD
BROYLES ROD		KEFJ	OPERATIONS CHIEF	RETIRED	NPS
BRYANT CAROL	ADMIN TECH	ANCH	PROCUREMENT	CACA	NPS
BUDGE CHUCK		ANCH	LOGISTICS SECTION CHIEF	RETIRED	NPS
BURCH JOHN	BIO TECH	ANCH	BEAR RESEARCH	DENA	NPS
BURGESS KEITH	CLERK TYPIST	ANCH	CLERK TYPIST	ARO	NPS
BUTLER CLAY		KEFJ	RPO	OLYM	NPS
BUTTERWORTH STEVEN		KEFJ	IIO TRAINER	PNR	NPS
CABANISS LOLA	ADMIN TECH	KEFJ	ADMIN.	KEFJ	NPS

Employee Name		Position	Location	Responsibility	Park	Agency
CABLE	JAY	CHIEF RANGER	KEFJ	ICS TRAINEE	KLGO	NPS
CANTRELL	BUD		KEFJ	RPO	BLRI	NPS
CARR	LAWRENCE		KATM	RPO	SEKI	NPS
CARTER	ALEX	CHIEF, RES. ASSES. BRANCH	KATM	BIRD SPECIALIST	ARO	NPS
CASE	JERRY	PARK RANGER/BIO TECH	KATM	PARK RANGER/BIO TECH	ISRO	NPS
CASEBEER	LOREN		KATM	RPO	FLETC	NPS
CASTELLINA	ANNE	SUPT, KEFJ	KEFJ	PARK MGMT	KEFJ	NPS
CAYOU	JOE	OPERATIONS	KATM	OPERATIONS/PLANS CHIEF	VOYA	NPS
CELLA	BRAD	RESOURCE SPECIALIST	KEFJ	RESOURCE SPECIALIST	ARO	NPS
CHISDOCK	TOM		KEFJ	RPO	ASIS	NPS
CLARK	DEAN	PARK RANGER	KATM	OPERATIONS	LAVO	NPS
CLARK	GLENN	CHIEF, INTERPRETATION DIV.	ANCH	PUBLIC RELATIONS	ARO	NPS
CLAWSON	LYNN	SUPPORT DISPATCH	KEFJ	DISPATCH		BLM/AFS
COATNEY	GAREY	CHIEF, LAND RESOURCES	KEFJ	INCIDENT COMMANDER	ARO	NPS
COE	KEN	DIV GROUP SUPERVISOR	KATM	DIV GROUP SUPERVISOR		BLM/AFS
COLLINS	BRUCE	PARK RANGER-BIO TECH	KATM	BIO TECH	GAAR	NPS
COOK	BILL		KATM	RES OPS RESOURCE	FLTEC	NPS
COOKE	GARY		KEFJ	LOGISTICS SECTION CHIEF		BIA
COPEMAN	ELIZABETH	CLERK TYPIST	KATM	CLERK TYPIST	KATM	NPS
COWAN	PETE	TOR CLAIM SPECIALIST	KATM	TORT INVESTIGATOR	GRCA	NPS
COWAN	PAUL		KEFJ	RPO	ARCH	NPS
COX	KAREN	PROGRAM ASSIT.	ANCH	RANGERS	ARO	NPS
COX	SHANNON	PROCUREMENT SPECIALIST	ANCH	PROCUREMENT	RMR	NPS
CROLL	MARGARET		ANCH	TRAVEL SPECIALIST	PNR	NPS
CROLL	STU	IC	KATM	IC TEAM	ISRO	NPS
CROUSSER	AL	SUPPLY LDR	KEFJ	SUPPLIES		USFS
CUMMINS	GARY T.		KEFJ	INCIDENT COMMANDER	CABR	NPS
CUSICK	JOEL		KATM	BIO TECH	KATM	NPS
DAPKUS	DAVID		LACL	SAFETY OFFICER		USFWS
DASH	DAVID		LACL	INCIDENT COMMANDER		BLM/AFS
DAVES	JAMA		ANCH	CONTRACTING SPECIALIST	BIBE	NPS
DAVIDSON	CATHY	INDUSTRIAL HYGENIST	ANCH	INDUSTRIAL HYGENIST	WASO	NPS
DAVIS	FRANCES		KATM	INCIDENT DISPATCHER		BIA
DAVIS	STEVE		KEFJ	RPO	FRED	NPS
DAWSON	RICK	BIO ADVIS.	KATM	BIO. ADVIS.	SER	NPS
DAWSON	RUTH		ANCH	SECRETARY	ARO	NPS
DAY	BRYAN	VIDEO EDITOR	KEFJ	VIDEO EDITOR		BIFC

Employee Name	Position	Location	Responsibility	Park	Agency
DEAN FRANK	TORT CLAIM SPECIALST	KATM	TORT INVESTIGATOR	YOSE	NPS
DENTON MEL	.	KATM	TORT INVESTIGATOR	GRTE	NPS
DERRICKSON JIM		ANCH	LABORER		AD
DESCHU NANCY	HYDRAULIC ENGINEER	KATM	WATER QUALITY	ARO	NPS
DEWITZ SCOTT	FILM CREW	KEFJ	FILM CREW		BLM/AFS
DICKENSON BOB	EQUIP MANAGER-BOATS	KATM	EQUIP MANAGER-BOATS		BLM/AFS
DIES DIXIE	ICT	KEFJ	IC TEAM		USFS
DILL PHIL		KEFJ	OPERATIONS CHIEF		BLM
DOOLAN CORY	SIT UNIT LEADER	KEFJ	SIT UNIT LEADER		BLM/AFS
DRAGOO DON	BEACH SURVEY SPEC.	KEFJ	BEACH SURVEY		USFWS
DROLET STEVE		KATM	RPO	LAME	NPS
DUGGINS DAVE	BIOLOGIST	KEFJ	MARINE BIOLOGIST		U OF WASH
DUNN BOB		ANCH	COMPUTER SPECIALIST	ARO	NPS
DUSTON REED	PARK RANGER	KEFJ	COASTAL RANGER	KEFJ	NPS
EALIES GLORIA	PROCUREMENT SPECIALIST	ANCH	PROCUREMENT	RMR	NPS
EASTWOOD JIM	SAFETY MGM	ANCH	RANGERS	ARO	NPS
EGAN LLOYD	COMPUTER PROGRAMMER	KEFJ	COMPUTER PROGRAMMER		BLM/AFS
ELIASON ALAN	SUPT., NWA	KEFJ	ICS TRAINEE	NWA	NPS
ELY GREG	RADIO TECH	KEFJ	RADIO		BLM/AFS
ERICKSON JON	PIO	KEFJ	PUBLIC INFORMATION	HAVO	NPS
ERSKINE DOUG	FIRE MANAGEMENT OFFICER	KEFJ	REGIONAL REPRESENTATIVE	ARO	NPS
ERSKINE CURT		ANCH	DISPATCH		AD
EVISON BOYD	REGIONAL DIRECTOR	ANCH	MANAGEMENT	ARO	NPS
FARO JAMES					ST. OF AK.
FAUROT DAVE		KEFJ	BIOLOGIST		USFWS
FEDOSH ROBERT		ANCH	ELECTRONICS TECH		BLM/AFS
FENNER ANDREA		KATM	CLERK		AD
FERTIG JOHN		KEFJ	RES. UNIT LEADER		USFS
FIBRANZ LYNN	SECRETARY	KEFJ	DATA ENTRY	ARO	NPS
FIELDS LUCY	TRAVEL CLERK	ANCH	TRAVEL	ARO	NPS
FINK WILLIAM	SUPERINTENDENT	KATM	TORT INVESTIGATOR	FONE	NPS
FINN JIM	WATER QUALITY/FISHERIES	KATM	WATER QUALITY/FISHERIES		USFWS
FIT ELAINE	COST ANALYST	KEFJ	COST ANALYST		USFS
FITZGERALD JACK	TORT CLAIMS	LACL	TORT INVESTIGATOR	CHIS	NPS
FITZMAURICE PETER	SUP RANGER, KEFJ	KEFJ	SUPERVISOR	KEFJ	NPS
FITZMAURICE ELAINE		KEFJ	PUBLIC INFO. OFFICER		USFS
FORBES MARK	RESOURCE USE SPECIALIST	KEFJ	RESOURCE USE SPECIALIST	PNR	NPS

Employee Name	Position	Location	Responsibility	Park	Agency
FOREMAN JANNA		KEFJ	MAIL DELIVERY	ARO	NPS
FORST RICHARD		KATM	RPO	SITK	NPS
FOWLER VELVA	MAIL & FILE CLERK	ANCH	SUPPLIES	ARO	NPS
FOWLER DAVE		KEFJ	RPO	EVER	NPS
FOWLER JOE		KATM	RPO	YELL	NPS
FRANKLIN MARK	R & D MGR	KEFJ	R & D MANAGER		USFS
FRAZIER BILL	TORT CLAIMS	LACL	TORT INVESTIGATOR	OLYM	NPS
FULLER DON	ICT	KEFJ	IC TEAM		BLM/AFS
FUTRELL JOE	PROCUREMENT CLERK	ANCH	PROCUREMENT	ARO	NPS
GABRIELSON PAUL	PHYCOLOGIST	KEFJ	PHYCOLOGY		U OF B.C.
GALE MARY ELIZABETH		ANCH	FINANCE SECTION CHIEF	GRCA	NPS
GAMET CAROL	ADMIN. TECH.	KATM	ADMINISTRATION	NEPE	NPS
GASPARINI STEVEN	SUPPORT DISPATCHER	ANCH	DISPATCH		ADOF
GERHARD BOB	RECREATIONAL SPEC.	LACL	RECREATIONAL SPEC.	LACL	NPS
GILBERT CHUCK	REALTY SPECIALIST	ANCH	BIOLOGIST (BIRD)	ARO	NPS
GLASS MIKE		KEFJ	PLANS SEC CHIEF	BADL	NPS
GLEESON PAUL	ARCHEOLOGIST	ANCH	ARCHEOLOGIST	ARO	NPS
GLEN TIANA	VIDEO EDITOR	KEFJ	VIDEO EDITOR		BIFC
GOHEEN TOM	ICT	LACL	IC TEAM		BLM/AFS
GORDON LOIS		KEFJ	FINANCE CHIEF	VOYA	NPS
GRAHAM LARRY		KEFJ	DRIVER		AD
GREENE LISA	BUDGET ANALYST	KATM	BUDGET	ARO	NPS
GREFFENIUS LAURA		KEFJ	COASTAL RANGER	KEFJ	NPS
GRIFFIN GENE	ARCHEOLOGIST	KEFJ	ARCHEOLOGY	ARO	NPS
GRIFFITHS LYNN		KEFJ	MAPPING SPECIALIST		NPS
GROSSMAN DARRELL	TORT CLAIM SPECIALIST	KATM	TORT INVESTIGATOR	ROMO	NPS
GRUBB JERRY		KEFJ	RPO	GUIS	NPS
GRZEGOROWICZ KAREN	PROCUREMENT SPECIALIST	ANCH	PROCUREMENT		NPS
GULLICKSON DAN		KEFJ	FILM CREW	SAGA	BLM/AFS
GULVESON DAVE	FILM CREW	KEFJ	FILM CREW		BLM/AFS
GUSTIN KAREN	RANGER	KEFJ	RANGER	KEFJ	NPS
HABSTER BILL	FUELER	KEFJ	FUELER		BLM
HAERTEL PAUL	ARD, RESOURCE SERVICES	ANCH	MANAGEMENT	ARO	NPS
HAMMOND JERRY	NPS RETIRED	KATM	SUPERINTENDENT'S REP.	RETIRED	RETIRED
HAMSON DAN	ENVIRONMENTAL SPECIALIST	ANCH	PLANS	ARO	NPS
HANABLE BILL	HISTORIAN	KEFJ	ADMIN HIST OF SPILL	ARO	NPS
HANNEMAN LARRY		KATM	RPO	LAME	NPS

Employee Name		Position	Location	Responsibility	Park	Agency
HARGER	BARBARA	SECRETARY	ANCH	SECRETARY	ARO	NPS
HARPHAM	D'LYN	CLERK TYPIST	KEFJ	CLERK TYPIST	KEFJ	AD
HARRIS	RICHARD	RESOURCE MANGEMENT SPEC.	KATM	BIOLOGIST (BIRDS)	BELA	NPS
HART	LESLIE	CHIEF, CULTURAL RESOURCES	ANCH	CULTURAL RESOURCES	ARO	NPS
HARVEY	MARK		KEFJ	RPO	LIBO	NPS
HATHAWAY	MARCUS	BUDGET ANALYST	ANCH	BUDGET	ARO	NPS
HAWKINS	CAT	WATER QUALITY SPEC.	KATM	WATER QUALITY	OLYM	NPS
HEACOX	KIM	BIRD BIOLOGIST	KATM	BIRD BIOLOGIST	KATM	AD
HEACOX	MELANIE	BIOLOGIST	KATM	MAMMAL BIOLOGIST	AAPLIC	NPS
HEAD	PAUL		LACL	PLANS SECTION CHIEF	LACL	BLM/AFS
HECKMAN	PHILIP		KATM	FINANCE CHIEF	GRCA	NPS
HELM	DOT	TERRESTRIAL VEG. BIOL.	KATM	TERRESTRIAL VEG. BIOL.	KATM	U OF A
HENDRIX	GARY	BIOLOGIST	ANCH	SCIENCE ADVISOR TO AC	SER	NPS
HENRY	LANA		ANCH	PROCUREMENT SPEC.	GEWA	NPS
HEPWORTH	JOHN		KATM	SIT/RES UNIT LEADER	KATM	USFS
HERBOLD	BONNIE		KEFJ	BIOLOGIST	KEFJ	AD
HERENDEEN	HEIDI		KATM	BIO TECH	KATM	NPS
HERMANNS	SHERRY		ANCH	PAYMENT TEAM	SAMO	NPS
HERRON	GEORGE		KEFJ	RPO	KEFJ	NPS
HEWSTON	SANDRA		ANCH	ADO PAYMENT TEAM	NATR	NPS
HEYT	KEN		LACL	SAFETY	LAVO	BLM
HINES	MEL	RADIO TECH	KATM	COMMUNICATION SPEC.	ARO	NPS
HOODENBACH	LOIS		ANCH	SAFETY	SWR	NPS
HOFFMAN	ROGER	BIRD BIOLOGIST	KATM	BIRD BIOLOGIST	OLYM	NPS
HOGAN	JOEL		KEFJ	RPO	DINO	NPS
HOLDA	WILLIAM		KATM	RPO	KATM	NPS
HOLDER	C.R.	ICT	KATM	IC TEAM	GRTE	BLM/AFS
HOLDER	STEVE	PARK RANGER	ANCH	LOGISTICS	JECA	NPS
HOLLAND	MARILYN		KEFJ		KEFJ	AD
HOLM	CHUCK		KEFJ	COMM. TECH	YELL	NPS
HOPKINS	JOE		ANCH	COMMUNICATION SPEC.	ARO	NPS
HOPSIER	WILLIAM		KEFJ	FUELER		BLM/AFS
HOUSTON	DOUGLAS	RESEARCH BIOLOGIST	KATM	RESEARCH BIOLOGIST	OLYM	NPS
HOWARTH	GINA		KEFJ	SECRETARY		AD
HUETHER	MARCIA		ANCH	CONTRACTING SPECIALIST	BADL	NPS
HUGHES	JACK	PARK RANGER	KATM	RPO	OLYM	NPS
HUMMEL	JIM	RESOURCE PROTECTION SPEC.	KATM	RPO	WRST	NPS

Employee Name		Position	Location	Responsibility	Park	Agency
HUNT	STEVEN	ENVIRONMENTAL SPECIALIST	ANCH	ENVIRONMENT	ARO	NPS
HUNTER	PAUL		KEFJ	ICS TRAINING	ARO	NPS
HURD	STEVE	PARK RANGER	KATM	RANGER	KATM	NPS
HUTCHISON	ANDY	SUPERINTENDENT, LACL	LACL	SUPERINTENDENT	LACL	NPS
ISAAC	JAKE		KATM	RPO	KATM	AD
JACKSON	JANA		KATM	DISPATCH	KATM	AD
JAMES	VIRGIL (RED)		KATM	RPO	CODA	NPS
JENSEN	MARVIN	SUPT, GLBA	KEFJ	ICS TRAINING	GLBA	NPS
JEWELL	LEE U.		KEFJ	FINANCE CHIEF	LIBO	NPS
JOHNNIE	ANDREW	SUPPLY CLERK	ANCH	SUPPLIES	ARO	NPS
JOHNS	THERESA	TRAVEL SPECIALIST	ANCH	TRAVEL	GLBA	NPS
JOHNSON	STEVE	SUPPORT DISPATCHER	KATM	DISPATCH	PNR	BLM/AFS
JOHNSON	DARRYL	RECREATION VALUES	ANCH	RECREATION VALUES	CUGA	NPS
JOHNSON	JIMMY		KEFJ	RPO	CUGA	NPS
JOHNSON	JOE		KATM	RPO	OZAR	NPS
JOHNSON	KYLE		KATM	RPO	GLAC	NPS
JONES	MARK		ANCH	DISPATCH (ORIENTATION)	GLAC	BLM/AFS
JORDAN	DICK	ARCHEOLOGIST	KATM	ARCHEOLOGIST		U OF A
JOY	DIANE	ADMIN. TECH.	KATM	ADMINISTRATION	SAJU	NPS
JUSTICE	KATHY		ANCH	COMPUTER SPECIALIST	ARO	NPS
KAISER	REBECCA	CHIEF, CONCESSIONS DIV.	KEFJ	RPO	ARO	NPS
KAMBITCH	JOHN	FILM CREW	KEFJ	FILM CREW		BIFC
KARRAKER	JEFF	PARK RANGER	KEFJ	PARK RANGER	YUCH	NPS
KARRAKER	DEAN	CONTRACT SPECIALIST	ANCH	CONTRACT SPECIALIST	MACA	NPS
KAVANAGH	ROSS	FISHERY BIOLOGIST	KATM	FISH BIOLOGIST	ARO	NPS
KELLEY	KEN		KATM	RPO	LAVO	NPS
KELLEY	ISAAC		KEFJ	RPO	PETE	NPS
KELLIHER	MARK	AD HIRE	ANCH	TRANSPORTATION		AD
KELSO	DONNA		ANCH	FINANCE CHIEF	ROMO	NPS
KEMPER	SUSAN		KEFJ	RPO	GLAC	NPS
KENNEDY	MARGARET		KEFJ	ASST. BIOLOGIST		AD
KERRIGAN	DONNA	PERSONNEL	KEFJ	PERSONNEL RECORDER		USFS
KING	RANDY		KEFJ	RPO	YELL	NPS
KING	NEIL		KEFJ	RPO	CRMO	NPS
KIRK	BILL	TERRESTRIAL VEG. BIO	KATM	VEG. BIOLOGIST		USFWS
KNAPP	KIP		KATM	RPO	JOTR	NPS
KNECHT	RICK	ARCHEOLOGIST	KATM	ARCHEOLOGIST	ARO	NPS

Employee Name	Position	Location	Responsibility	Park	Agency
KNEIPP GREGG		KATM	RPO	NCR	NPS
KNIPPER CAROL	RPO	KATM	RPO	JODA	NPS
KNOWLES RON	ICT	KEFJ	IC TEAM		USFS
KNUCKLES DENNIS	RANGER	LACL	TORT INVESTIGATOR	YUCH	NPS
KNUDSON ROBERT		KEFJ	DRIVER		AD
KORTGE LLOYD		KATM	OPERATIONS SECTION CHIEF	BADL	NPS
KRAUSHAAR JOHN	PARK RANGER	ANCH	AREA COMMANDER	SEKC	NPS
KRUMENAKER ROBERT	BIO TECH	KATM	BIO TECH	ISRO	NPS
KUCINSKI RUSS		KEFJ	INCIDENT COMMANDER	ARO	NPS
KYLE SCOTT		KATM	AIR OBSERVER		AD
LADD BENJAMIN F.	SUPERINTENDENT	KATM	RPO	JODA	NPS
LALONE MICHAEL		KEFJ	RPO	YOSE	NPS
LATTEREL ART	TRAINING OFFICER	KEFJ	TRAINING		BLM/AFS
LAUGHLIN KAYE	RANGER	ANCH	RANGER	ARO	NPS
LAWRENCE BILL	ENV COMPLIANCE	ANCH	ENV COMPLIANCE	ARO	NPS
LAWSON LINDY	SECRETARY	KEFJ	SECRETARY	ARO	NPS
LAWSON HAL	COMPUTER SPECIALIST	KEFJ	COMPUTER SPECIALIST	ARO	NPS
LEACH HOMER		KEFJ	TORT INVESTIGATOR		AD
LEE NORMAN	CHIEF APPRAISER	KEFJ	TECH. SPECIALIST	ARO	NPS
LEE LOGAN	RES UNIT LEADER	KATM	RES UNIT LEADER		USFS
LENTFER HENRY	MARINE ECOLOGIST	KATM	MARINE ECOLOGY	GLBA	NPS
LEWIS JACK	EQUIP MGR. - BOATS	KEFJ	BOAT EQUIP. MGMT.		BLM/AFS
LIEBERSBACH DAVE	ICT LEADER	KEFJ	INCIDENT COMMANDER		BLM/AFS
LIEN LINDSEY	SUPPORT DISPATCH	ANCH	DISPATCH		BLM/AFS
LINDERMAN LINDA	SECRETARY	KEFJ	FINANCE CHIEF	MWR	NPS
LINDSAY BOB		KEFJ	SIT UNIT LEADER		USFS
LINK KRISTI LEE		KEFJ	COASTAL RANGER	KEFJ	NPS
LITTLE MARK		KEFJ	BIOLOGIST		AD
LOGAN CHARLES		KEFJ	OPERATIONS CHIEF	GLAC	NPS
LOVAAS AL	SCIENTIST	ANCH	SCIENTIST	ARO	NPS
LOWIN DONNA	ADMIN. OFFICER	ANCH	PROCUREMENT	BIBE	NPS
LUNDSFORD JERRY		KEFJ	RPO	KEFJ	NPS
LYNCH DAVID		KEFJ	LABORER		AD
MAGGIORA MARK		KEFJ	SIT. UNIT LEADER		USFS
MANSKI DAVE	BIO ADVISOR	KATM	BIOLOGIST	KATM	NPS
MARTIN MARY	TRAINING OFFICER	ANCH	TRAINING	ARO	NPS
MARTIN CHRIS		KATM	BIO TECH	KATM	NPS

Employee Name	Position	Location	Responsibility	Park	Agency
MASON MARVIN	HELIBASE MGR.	KATM	HELIBASE MANAGER		USFS
MATT COLLEEN	COASTLINE AERIAL RES. SPEC.	KATM	COASTLINE AERIAL RES.SPEC		AF&G
MCCLENAHAN PATRICIA	ARCHEOLOGIST	KATM	ARCHEOLOGIST	ARO	NPS
MCCREIGHT ROCKY		KEFJ	RPO	GRTE	NPS
MCGUINESS SEAN		KEFJ	RPO	CRLA	NPS
MCKEEMAN BRUCE		KATM	TORT INVESTIGATOR	GRFA	NPS
MCKNIGHT REX	FIXED WING BASE MGR.	KEFJ	FIXED WING MANAGER		BLM/AFS
MCMANUS DICK		KEFJ	IC TEAM		BLM/AFS
MCWILLIAMS LOREN		KEFJ	BARRACKS MANAGER		AD
MEARS DON	TIME UNIT READER	KATM	TIME UNIT READER		BLM/AFS
MEEHAN JOSEPH		KEFJ	COASTAL RANGER	KEFJ	NPS
MELDRUM JANIS	RESOURCE MGMT	KATM	RESOURCE MGMT	KATM	NPS
MEYER JOHN	FISHERIES SPECIALIST	KATM	FISHERIES SPECIALIST	OLYM	NPS
MICHAELSON JULIE	BIOLOGIST	KEFJ	VEGETATION SPECIALIST	ARO	NPS
MICHELS BILL		KEFJ	RPO	GLAC	NPS
MILLER KATHY ANN	BIOLOGIST	KEFJ	MARINE BIOLOGIST		U OF WASH
MILLER ANNE		KEFJ	TECH SPECIALIST		AD
MILLER ERIC		KATM	HELIBASE MANAGER		BLM/AFS
MILLER JOHN		KATM	RPO	SWR	NPS
MILLER BILL	PARK RANGER	KATM	SUPERINTENDENT'S REP		AD
MILLS DAVE		LACL	TORT INVESTIGATOR	NWA	NPS
MILNER SANDY	BIOLOGIST	KEFJ	FISH BIOLOGIST		AD
MILSTEIN MICHAEL		KATM	RPO	DETO	NPS
MITCHELL SUE	IIO	KATM	IIO		BLM/FSC
MOORE ZACHARY		KEFJ	RPO	WR	NPS
MOREFIELD RICHARD		KEFJ	RPO	BLRI	NPS
MORTON TOM	PARK RANGER	KATM	RPO	YOSE	NPS
MOSELEY MARK		KATM	RPO	BUNA	NPS
MOW JEFF		ANCH	TORT INVESTIGATIONS	ARO	NPS
MULDOON CICELY	LEGAL DATA CATALOGER	KATM	LEGAL DATA CATALOGER	SITK	NPS
MURDOCK IDA	RANGER, KEFJ	KEFJ	RANGER	KEFJ	NPS
MYERS JOHN	CARTOGRAPHIC TECHNICIAN	KEFJ	CARTOGRAPHY	ARO	NPS
NELSON BENJAMIN		KEFJ	RPO	PNR	NPS
NELSON EDWARD E.		KATM	AIR SUPPORT SUPERVISOR	SEKI	NPS
NEMETH DAVID		KATM	TRAINING DIRECTOR	KATM	NPS
NICHOLS GREG	COMM. TECH	KATM	ICP		BIFC
NISHIMOTO MIKE	BIOLOGIST	KEFJ	BIRD BIOLOGIST		USFWS

Employee Name		Position	Location	Responsibility	Park	Agency
O'CONNEL	TERRY	LOGISTICS CHIEF	LACL	LOGISTICS		BLM/AFS
O'DANIEL	MARY JANE	EQUIP/TIME RECORDER	KATM	EQUIP/TIME RECORDER		BLM/FSC
O'DEA	JACK	DISTRIBUTION	KATM	DISTRIBUTION		BIA
O'GUIN	RICH	CHIEF PROT. & RANGER ACTIV.	ANCH	RANGERS	ARO	NPS
OELFKE	JACK	PARK RANGER	KATM	RPO		NPS
OLDOW	DEBBIE		KEFJ	CLERK TYPIST		AD
OLIVER	ROY	PAYMENT TEAM	KEFJ	PAYMENT TEAM		BLM
OLSON	GORDON		KATM	RESEARCH BIOLOGIST	ASIS	NPS
ORADEI	DAVID		ANCH	MAPPING	ARO	NPS
ORLANDO	CYNDY		KEFJ	TRAINER	PNR	NPS
OROT	SALLY		LACL	ADMIN TECH	LACL	NPS
ORR	BILL		ANCH	LOGISTICS SECTION CHIEF	RETIRED	NPS
OSWALDT	DAVE	DISPATCH	KEFJ	DISPATCH		BLM/AFS
OVERTON	HOWARD		KATM	RPO	CABR	NPS
PACE	GARY		KEFJ	RPO	CUYA	NPS
PAGE	SUZY	SECRETARY	ANCH	SECRETARY	ARO	NPS
PARKER	GENE		KATM	RPO	BLRI	NPS
PARKES	SEYMOUR		KEFJ	RPO	OLYM	NPS
PATTERSON	RALPH		KATM	RPO	LAME	NPS
PAUL	PETER	LOGISITICS & SUPPLY	KEFJ	LOGISITICS & SUPPLY		AD
PAULUS	KEVIN	STILL PHOTOGRAPHY	KATM	PHOTOGRAPHY		AD
PAYER	DAVE		KATM	BIO TECH	KATM	NPS
PEARSON	CHRIS	ORDERING MANAGER	KATM	ORDERING MANAGER		BLM/AFS
PENTTILA	TERRY		KEFJ	RPO	RMR	NPS
PETERSON	JERRY	R & D MANAGER	KATM	R & D MANAGER		BLM/AFS
PETERSON	JOHN		KATM	RPO	GRCA	NPS
PHELAN	PAT	CHIEF, BUDGET	ANCH	BUDGET	ARO	NPS
PILLSBURY	VALERIE		ANCH	PAYMENT TEAM	LAVO	NPS
PIORKOWSKI	ROBERT		KEFJ	WATER QUALITY ASST.		AD
POLLOCK	KEITH		KEFJ	RADIO TECH		BLM/AFS
PONTBRIAND	EDWARD		KATM	RPO	WICA	NPS
PONTBRIAND	DANIEL	PARK RANGER	KATM	RPO	BICA	NPS
POOLE	JAMES		KATM	RESEARCH BIOLOGIST	NCR	NPS
PURIFOY	PAUL		KEFJ	RPO	EVER	NPS
QUINLEY	JOHN	PUB AFFAIR	ANCH	PUB AFFAIR	ARO	NPS
RABINOWITCH	SANDY		KEFJ	ICS TRAINEE	ARO	NPS
RADER	JEFF		KATM	TORT INVESTIGATOR	GRTE	NPS

Employee Name		Position	Location	Responsibility	Park	Agency
RAMBO	WOODY		KATM	RPO	OLYM	NPS
RANDALL	JIM		ANCH	PLANS CHIEF	RETIRED	NPS
RANDALL	ROBERT	R & D MGR	KATM	RPO	CABR	NPS
REED	TIM		KATM	MANAGEMENT		BLM/AFS
REED	HARRY		KEFJ	PETROLEUM OBSERVER		ADEC
RIBAR	JOE	ICT	KEFJ	IC TEAM		BLM/AFS
RICE	WILLIAM	RES MGMT,KEFJ	KEFJ	RES MGMT	KEFJ	NPS
RICHARDSON	JOHN		KEFJ	BOOMER GROUP		AD
RICHTER	PETER		KEFJ	ICS TRAINEE	ARO	NPS
RIGBY	WARREN		KEFJ	ICS TRAINEE	KOVA	NPS
RILEY	JIM		KATM	RPO	LAME	NPS
RITCHIE	BRENDA	DISPATCH	KATM	DISPATCH	SHEN	NPS
RITCHIE	BOYD	LOGISTICS CHIEF	KATM	LOGISTICS CHIEF		USFS
ROBERTSON	MARV	ICT	KEFJ	IC TEAM		AK/DOF
ROBINSON	BEN	ASSIT.LOG. CHIEF	LACL	ASSIST. LOG. CHIEF		BLM/AFS
ROBINSON	STEVE		KEFJ	RPO	MEVE	NPS
ROESSLER	JIM		LACL	OPERATION SECTION CHIEF		BIA
ROGERS	STACEY	AK STAGING AREA MGR	KATM	MANAGEMENT		BLM/AFS
RONDAS	MICHAEL		KEFJ	RPO	LAME	NPS
RONEY	KATE	PARK RANGER/BIO TECH	KATM	PARK RANGER/BIO TECH	NWA	NPS
ROOS	MIKE	FILM CREW	KEFJ	FILM CREW		BLM/AFS
ROSENBERG	TOM	HELIBASE MANAGER	KATM	HELIBASE MANAGER		USFS
ROSSINI	BETSY	ADMIN. PAYMENT SPEC.	KEFJ	ADMIN PAYMENT SPEC.	YUCH	NPS
ROY	CORDELL	ENVIRONMENTAL SPECIALIST	ANCH	PLANS	ARO	NPS
RUARK	DON		KEFJ	LOGISTICS SECTION CHIEF		USFS
RUMMELE	LAURA		KATM	ADMINISTRATION	KATM	NPS
RYAN	JIM		ANCH	FINANCE	RETIRED	NPS
RYAN	CHRIS		KATM	RPO	JENA	NPS
SALO	LEANN	ADMIN ASSITANT	KEFJ	ADMIN ASSISTANCE		BLM/AFS
SAMORA	BARBARA	RESOURCE MGMT. SPECIALIST	KATM	RESOURCE MANAGEMENT	MORA	NPS
SAND	ERIC		KEFJ	BOOMER		AD
SANDERS	JOAN	ADMIN. TECH	KEFJ	ADMIN TECH	HATR	NPS
SAUNDERS	RICHARD		KATM	RPO	BOWA	NPS
SCHAFF	JEAN	ARCHEOLOGIST	KEFJ	ARCHEOLOGIST	ARO	NPS
SCHEIZSLE	TONY		KATM	IC TEAM	CANY	NPS
SCHLINKMANN	COLLETTE	SEASONAL	ANCH	SECRETARY/BUDGET ASST.	ROMO	NPS
SCHMIDT	RICHARD		KEFJ	LABORER		AD

Employee Name	Position	Location	Responsibility	Park	Agency
SCHOCH CARL	R & D HELPER	KEFJ	R & D HELPER		AD
SCHOENBERG KEN	ARCHEOLOGIST	ANCH	ARCHEOLOGIST	ARO	NPS
SCHREINER ED	BOTONIST	KATM	BOTONIST	OLYM	NPS
SCHROEDER MARK	CHIEF RESOURCE MANAGEMENT	KATM	RESOURCE MANAGEMENT	GLBA	NPS
SEBADE GARY		KATM	RPO	LAME	NPS
SELA MICHAEL		KEFJ	MAPPING CREW		AD
SHACKELTON STEVE	LAW ENFORCEMENT	ANCH	LAW ENFORCEMENT	ARO	NPS
SHACKELTON LEE	TORT CLAIM SPECIALST	KEFJ	TORT INVESTIGATOR	YOSE	NPS
SHAVER MACK		KEFJ	OPERATION SECTION CHIEF	THRO	NPS
SHAW BRUCE		KATM	RPO	RETIRED	NPS
SHEEHAM JOAN	AFS SPEC.	ANCH	AFS SPECIALIST	NAR	NPS
SHERMAN RICHARD	MTNCE WKR	KATM	MAINTENANCE	KATM	NPS
SHERMAN WILLIAM		KEFJ	RPO	LAME	NPS
SHUTE DIANE		ANCH	PROCUREMENT SPEC.	MORA	NPS
SIEBECKER ALICE	LIAISON OFFICER	KEFJ	LIAISON	YELL	NPS
SIKES C. NEWTON		KATM	INCIDENT COMMANDER	LAME	NPS
SMITH RON	IIO	KEFJ	IIO	KEFJ	BLM
SMITH AL	RESOURCE PROTECTION SPEC.	KATM	RPO	DENA	NPS
SMITH TIM	ARCHEOLOGIST	KEFJ	ARCHEOLOGIST	ARO	NPS
SMITH FRANK		KEFJ	RPO	FRED	NPS
SMITH JANELLE		ANCH	FINANCE	ARO	NPS
SMITH GEOFF		KEFJ	COASTAL RANGER	KEFJ	NPS
SNYDER HANK		KEFJ	RPO	GEWA	NPS
SORENSON HARVEY	RESOURCE LDR	KATM		WASO	NPS
SPARHAWK STEVE		KEFJ	RPO	CRMO	NPS
SPARKS DIXIE	PROCUREMENT	ANCH	PROCUREMENT/CONTRACTING	MEVE	NPS
SPECKMAN KIM	RANGER (PILOT)	KATM	RANGER (PILOT)	KATM	NPS
SPENCER PAGE	ENV SPEC.	ANCH	ENV SPECIST	ARO	NPS
SPIRTES DAVE	RANGER	KEFJ	ICS TRAINEE	GLBA	NPS
SPONSEL ART	CHIEF, PROCUREMENT	ANCH	PROCUREMENT	ARO	NPS
SPONSEL BRIAN	AD HIRE	ANCH	TRANSPORTATION		AD
SQUIBB RON	RESOURCE MANAGEMENT SPECIALIST	KATM	COASTLINE AERIAL RECON	KATM	NPS
STAM JOE	ICT	KEFJ	IC TEAM		AK/DOF
STANSBERRY SALLY		ANCH	CONTRACTING SPECIALIST	MTRA	NPS
STENMARK DICK	DEPUTY REGIONAL DIRECTOR	ANCH	ADMINISTRATION	ARO	NPS
STEVENS WILLIAM	MTNCE WKR	KEFJ	MAINTENANCE	KEFJ	NPS
STEVENS SILUS	R & D MANAGER	LACL	MANAGER		BLM/AFS

Employee Name		Position	Location	Responsibility	Park	Agency
STEVENS	DAVID		KATM	RESEARCH BIOLOGIST	ROMO	NPS
STILIPEC	ROGER	DISPATCH	KATM	DISPATCH	KATM	BLM/AFS
STINGLEY	SUSIE		ANCH	DISPATCH		BLM/AFS
STOMBACK	JANET		KATM	DISPATCH	SHEN	NPS
STONDALL	ED	MTNCE MECHANIC	KATM	MTNCE MECHANIC	KATM	NPS
STONE	TIM		KATM	RPO	GOGA	NPS
STONE	ROGER		KEFJ	RPO	HOFU	NPS
STRAND	RICH	COST UNIT TEAM	KATM	COST UNIT TEAM		USFS
STROBE	ROBERT		KEFJ	ICS TRAINEE	ARO	NPS
STROMME	PHYLLIS	PURCHASING AGENT	ANCH	PROCUREMENT	ARO	NPS
STRUNK	DON		KATM	LOGISTICS		USFS
SUMMERFIELD	JUDY	TRAVEL SPECIALIST	ANCH	TRAVEL	ROMO	NPS
SUMMERS	CLARENCE	SUBSISTENCE SPECIALIST	ANCH	SUBSISTENCE	ARO	NPS
SUTTON	LARRY		LACL	DISPATCH		BLM/AFS
SWAIN	TODD	PARK RANGER	KATM	RPO	JOTR	NPS
SWIFT	KATHERINE		KATM	BIO TECH	KATM	NPS
SYPHER	CHUCK		KATM	FIELD OBSERVER	LAVO	NPS
TALSMA	CARL		KATM	RPO	GLAC	NPS
TAYLOR	DALE	BIOLOGIST	KEFJ	BIOLOGIST	ARO	NPS
TAYLOR	SCOTT	TORT CLAIM SPECIALST	KEFJ	TORT INVESTIGATOR	SITK	NPS
TENNESON	RENE	DISPATCH	KEFJ	DISPATCH		BLM/AFS
TETREAU	MIKE	VIP	KEFJ	TERRESTIAL ECOLOGIST	KEFJ	NPS
THATCHER	ROBERT		KEFJ	RPO	VICK	NPS
THOMAS	DIANA	RANGER	KEFJ	RANGER	KEFJ	NPS
THOMAS	JOHN	SUP.LOGISTICS DISPATCH	KEFJ	SUP.LOG. DISPATCH		BLM/AFS
THOMPSON	DONNA		KEFJ	CLERK		AD
THORPE	CARL		KEFJ	AIRCRAFT FUELER		BLM/AFS
TIECKE	CLARK		KATM	HELIBASE MANAGER		BLM
TOMS	LINDA		KATM	FINANCE CHIEF	CHOH	NPS
TROYER	WILL		KATM	BIO TECH	RETIRED	NPS
TSCHOHL	THOMAS	SUPV. PARK RANGER	KATM	PLANS/OPERATIONS CHIEF	SEKI	NPS
TURNER	GORDON		KEFJ	DRIVER		AD
TWEED	WILLIAM		KATM	INFORMATION OFFICER	SEQU	NPS
TWITCHELL	HOLLIS	PARK RANGER/PILOT	LACL	GROUP SUPERVISOR	LACL	NPS
VALENTA	THOMAS		KATM	RPO	LAME	NPS
VALLIER	GLORIA		ANCH	PROCUREMENT	RMR	NPS
VAN ALSTINE	NANCY		KATM	RES MGMT SPEC	GAAR	NPS

Employee Name	Position	Location	Responsibility	Park	Agency
VAN SLYKE LARRY	SUPERVISORY PARK RANGER	LACL	RANGER	LACL	NPS
VANDERLINDEN LARRY		LACL	PLANNING SECTION CHIEF	LACL	USFWS
VEQUIST GARY	RES MGMT	KEFJ	RES MGMT	ARO	NPS
VINSON DALE		KEFJ	ARCHEOLOGIST	ARO	NPS
VONNER AL	PARK RANGER	KATM	RPO	CAMO	NPS
WAGERS WILLIAM		KEFJ	RPO	PEFO	NPS
WAGNER GEORGE	PARK RANGER	KATM	SIT UNIT LEADER	DENA	NPS
WAHL DON	ICT	KEFJ	IC TEAM		BLM/AFS
WALLER LOU	CHIEF, SUBSISTENCE	ANCH	ECONOMICS	ARO	NPS
WALTERS JIM		KEFJ	RPO	SWR	NPS
WARBURTON JANICE	MAP RECORDER	KEFJ	MAPPING	ARO	NPS
WARD JIM	ICT	KEFJ	IC TEAM		BLM/AFS
WARREN JUDITH	SECRETARY	KEFJ	SECRETARY		AD
WARREN RAY		KATM	AIR SUPPORT SUPERVISOR	RETIRED	NPS
WASKA ADAM		ANCH	COMMUNICATION SPEC.	ARO	NPS
WEATHERBY THOR	COMM. TECH	KEFJ	COMM. TECH		BLM/AFS
WEEMS LEONARD		KEFJ	RPO	SWR	NPS
WEGENER JOSEPH	PARK RANGER	KATM	RPO	LAME	NPS
WEHKING LEONARD	RESOURCE/SIT U.L.	KATM	RESOURCE/SIT U.L.		BLM/AFS
WEILAND DENNIS		KEFJ	RPO	YOSE	NPS
WEINS LYNN		KEFJ	RPO	SWR	NPS
WELCH BILL	MANAGEMENT ASSISTANT	ANCH	RESPONSE MANAGEMENT	ARO	NPS
WELCH JACOB		KEFJ			AD
WELLS JAY	RESOURCE MANAGER	KATM	SUPERINTENDENT'S REP.	WRST	NPS
WESTPHAL WAYNE		KEFJ	RPO	DEVA	NPS
WHEELER MARCELLA		ANCH	TRAVEL SPECIALIST	SWR	NPS
WHITE VICKIE		KEFJ	FINANCE SECTION CHIEF	MWR	NPS
WHITE MATTHEW		KEFJ			AD
WHITE ROBERT G.		KATM	INCIDENT COMMANDER	RETIRED	NPS
WHITEMAN ROBERT	RANGER	KATM	RPO	COLO	NPS
WHITMER GUY		KEFJ	RPO	LASS	NPS
WHYTE CLYDE		KATM	HELIBASE MANAGER	MEVE	NPS
WILLIAMS SHELLY	BIOLOGIST	KEFJ	FISH BIOLOGIST		AD
WILLIAMS BRUCE	RECEIVE/DISTRIB.	KEFJ	RECEIVING/DISTRIB.		USFS
WILLIAMS RAWLES		ANCH	INCIDENT DISPATCHER		AD
WILLIAMS JAN		KATM	LOGISTICS		AD
WILLIAMS M. "SCHELLE"		ANCH	PROCUREMENT	PEFO	NPS

Employee Name		Position	Location	Responsibility	Park	Agency
WILLIAMSON	LAURIE	SUPPORT DISPATCHER	KATM	DISPATCH		BLM/BIFC
WINTER	WAYNE		LACL	LOGISTICS SECTION CHIEF		USFS
WISLEY	DIANA	SUPPORT DISPATCHER	ANCH	DISPATCH	PNR	NPS
WITT	MARY	FINANCE SEC CHIEF II	KATM	FINANCE SECTION CHIEF II		BLM/AFS
WIZNER	NANCY		KATM	RPO	CAMO	NPS
WOLVERTON	DAVID	CONTRACTING SPECIALIST	ANCH	CONTRACTING SPECIALIST	ARO	NPS
WOOD	SHERRY	FINANCE CLERK	KATM	FINANCE CLERK	LAVO	NPS
WOODS	MIKE	HELICOPTER MGR.	KATM	HELICOPTER MGR		BLM/AFS
WORLEY	MIKE	DISPATCH	KEFJ	DISPATCH		BLM/AFS
WORTHINGTON	ANNE		KEFJ	ARCHEOLOGIST	ARO	NPS
WRIGHT	LARRY	ENV COMPLIANCE	ANCH	ENV COMPLIANCE	ARO	NPS
WRIGHT	SHERRY	SUPPLY CLERK	ANCH	SUPPLIES	ARO	NPS
YOUNG	BLAIR	SIT UNIT LEADER	KEFJ	SIT UNIT LEADER		BLM/AFS
YOUNGER	JOY		ANCH	PROCUREMENT	ARO	NPS
YURICK	MAGGIE		KATM	BIO TECH	KATM	NPS
ZWINGER	SUSAN		KEFJ	CLERK	KEFJ	NPS